A PRACTICAL GUIDE TO

ENVIRONMENTAL
IMPACT ASSESSMENT

A PRACTICAL GUIDE TO
ENVIRONMENTAL
IMPACT ASSESSMENT

PAUL A. ERICKSON

New England Research, Inc.
Worcester, Massachusetts

ACADEMIC PRESS

San Diego   New York   Boston   London   Sydney   Tokyo   Toronto

Copyright © 1994 by ACADEMIC PRESS, INC.

All Rights Reserved.
No part of this publication may be reproduced or transmitted in any form or by any
means, electronic or mechanical, including photocopy, recording, or any information
storage and retrieval system, without permission in writing from the publisher.

Some material previously published in: ENVIRONMENTAL IMPACT
ASSESSMENT, copyright © 1979 by Academic Press, Inc. All rights
reserved.

Academic Press, Inc.
A Division of Harcourt Brace & Company
525 B Street, Suite 1900, San Diego, California 92101-4495

United Kingdom Edition published by
Academic Press Limited
24–28 Oval Road, London NW1 7DX

Library of Congress Cataloging-in-Publication Data

Erickson, Paul A.
    A practical guide to environmental impact assessment / by Paul A.
Erickson.
        p.   cm.
    Includes bibliographical references and index.
    ISBN 0-12-241555-8
    1. Environmental policy. 2. Environmental impact analysis.
GE170.E75 1994
333.7'14--dc20   93-49572
                                                               CIP

PRINTED IN THE UNITED STATES OF AMERICA
        96 97 98 99   BB   9 8 7 6 5 4 3 2

For
Terry and John Reynolds,
teachers and friends, both.

CONTENTS

PRINCIPLES OF ENVIRONMENTAL IMPACT ASSESSMENT

THE PHYSICAL ENVIRONMENT

THE SOCIAL ENVIRONMENT

ISSUES OF SPECIAL CONCERN

PREFACE

In 1979, I authored a text on impact assessment (*Environmental Impact Assessment: Principles and Applications,* Academic Press). Its contents were based on the requirements of the U.S. National Environmental Policy Act and on what I considered to be the particular strengths and weaknesses of impact assessments that had been conducted in the United States in the near decade since the passage of this law.

Environmental impact assessment has now become a standard tool of decision making throughout the world—in both industrialized and developing nations. Our experience with impact assessment not only has deepened in terms of its continual practice and subsequent refinement, but also has expanded to a wide range of human and environmental contexts and conditions that, fourteen years ago, were hardly if ever considered.

There can be no doubt that the tools, techniques, and protocols available for the purpose of environmental impact assessment have been undergoing dramatic changes—changes that, if not instigated by, have at least been hastened by the continually developing advances in such technologies as remote sensing, computer-assisted data analysis and display, bioassays, and mathematical modeling.

Having so many new and increasingly sophisticated tools to apply to so many both long and newly recognized problems associated with human-mediated impacts on the environment might suggest that we are well beyond the formative stage of environmental impact assessment. Yet, it may also be reasonably suggested that the sophistication of analytical techniques does not necessarily translate into the sophistication of the final analysis—that having new tools at our disposal does not mean we know how best to use them to achieve the comprehensive objectives of environmental impact assessment.

This book is intended as a practical guide to the technical and scientific concepts that must be addressed in any comprehensive assessment of project-mediated impacts on the complex, interacting physical and social components, attributes, and dynamics of the environment. While specific analytical protocols and methodologies are identified when necessary or by way of example, the emphasis is clearly on conceptualization—on the identification and practical application of concepts that must guide any systems analysis of the biosphere.

I do not assume that the reader is professionally trained in any particular natural or social science discipline. Materials are presented in such a manner that the discerning reader may easily identify which areas of disciplinary knowledge should perhaps be consulted by additional reading of the relevant literature.

Because environmental impact assessment is a global activity and is therefore subject to diverse interests and concerns, I have ignored certain procedural or technical requirements of impact assessment that are peculiar to specific legislative or executive jurisdictions. In this sense, then, I have focused on how one identifies and evaluates impacts on the interdependent physical and social environment rather than on how one may, in the process of impact assessment, comply with a particular legal requirement.

With these considerations in mind, I intend this book to be a practical conceptual guide for those who have professional responsibility for the design, management, or conduct of impact assessment. I also believe that it will prove useful to the student who, regardless of disciplinary background, wants to examine in greater detail the kinds of interdisciplinary contributions to contemporary decision-making processes that characterize modern project development.

I am greatly indebted to Claire G. (Myers) Erickson, not only for her considerable editorial skills, but also for her important help in collecting and organizing the formidable amount of information that had to be considered in the development of this text.

PRINCIPLES OF ENVIRONMENTAL IMPACT ASSESSMENT

CHAPTER I

OVERVIEW OF ENVIRONMENTAL IMPACT ASSESSMENT

BACKGROUND

Environmental impact assessment is the process of identifying and evaluating the consequences of human actions on the environment and, when appropriate, mitigating those consequences. Precise elements of this process and specific technical guidelines and legal requirements that pertain to it vary from nation to nation and, within one nation, vary with time. After more than 20 years as a formally recognized tool of decision making, environmental impact assessment remains an evolving art and science, sensitive to a plethora of sometimes consonant but often conflicting local, regional, national, and global concerns and interests.

Despite the diversity of techniques, the differences in emphasis, and the varied objectives that characterize impact assessment as practiced in different nations, four important aspects of environmental impact assessment are increasingly approaching consensus.

First, and foremost in importance, is the fact that environmental impact assessment requires seeing the environment as **the aggregate of things and conditions that surround or envelop every living and nonliving thing.** From this perspective, the environment is not just the physical habitat of waterfowl, for example; it is also the water, soil, air, and plants that make up that habitat. The environment is not just the forest; it is also all the processes that take place in, around, and because of that forest. Nor is the environment merely those things and processes that are not human. **The environment also includes humans and the things, processes, and conditions that pertain to humans.**

During the American development of impact assessment, this view of the environment as the aggregate of all living and nonliving things and processes was denoted by such phrases as "the total environment" and "the human environment." More recently, individuals in this field have increasingly recognized that such phrases are needlessly redundant. Currently, "environment" simply means the physical, chemical, biological, and social entities, conditions, and dynamics that surround us. Environmental impact assessment is, therefore, the effort to (1) determine how our actions might change these entities, conditions, and dynamics; (2) establish criteria by which to evaluate the desirability of such changes; and (3) mitigate selected changes by appropriate engineering or management techniques.

Second, as a tool of decision making, the value of environmental impact assessment is more likely to be realized in the timely communication of information between individuals conducting the assessment and individuals planning a proposed project than in the writing of a massive technical document that few if any decision makers will ever read. Although written assessments, often called Environmental Impact Statements (EISs) or Environmental Impact Reports (EIRs), are required by legislative or executive mandate, our general understanding is that such documents tend to serve more as records of decision making than as active tools in decision making. The ideal is an iterative assessment process that begins at the earliest phase of project development and, running concurrently with project development, constitutes an integrated feedback loop with project planning, implementation, and operation.

Third, although many environmental components, processes, and attributes are amenable to currently available methods of quantification, many are not. By their nature, these factors may never become quantifiable, regardless of the continuing development of new analytical paradigms. Any attempt to restrict assessment efforts solely to quantifiable aspects of the environment or to numerical analysis is, therefore, a gross simplification of the actual environment and, therefore, an unacceptable input into decision making.

Fourth, mitigation of significant impacts, which includes the minimization of undesirable impacts and the enhancement of desirable impacts, must be assessed for all possible impacts. For example, developers may decide that borrow pits within a project area should be located and designed so they can be used as shallow impoundments and wetland areas after completion of the project. This approach may mitigate the loss of other wetland and wildlife habitat in the area caused by project location or design. However, all real-world consequences of such a decision must be considered. For instance, how will the new wetland habitat affect public health, since it can serve as habitat for mosquitos or other disease vectors? Will the new wetland attract waterfowl in such numbers that their fecal droppings into existing surface waters in the area will lead to water quality problems in potable supplies? What are the consequences for the physical safety of children in the area?

Examining these four aspects of impact assessment in greater detail is worthwhile, particularly because they directly influence our perception of the sources of impacts.

PHYSICAL AND SOCIAL ENVIRONMENTS

An appreciation of the environment as the total aggregate of living and nonliving things, conditions, and processes is precisely reflected in the bromide, "All things are connected to one another." We may intuitively grasp the sense of this statement but, in practical terms, we know that all things are not connected equally—that everything in the world does not change because one part does. We can reasonably expect that some components of the environment can change or be altered (even drastically) with no measurable effect on some other component. Although we accept the doctrine of holism—that the environment is a whole—practical experience tells us to discern the degree of relationship or connectedness among the individual parts of any whole. Practical experience also tells us that the identities of the individual parts of any whole may vary with the level of discrimination used, that is, with the way we view the whole and the way we choose to analyze it.

Technically, the various levels of discrimination used in the analysis of a complex system may be referred to as homomorphic models. One homomorphic model of a lake may focus on atoms and molecules. Another may focus on various categories of these components, for example, organic molecules. Still another homomorphic model may focus on entirely different components, including various species of invertebrate animals and selected parameters such as temperature and the concentration of oxygen, that may influence those species. Each homomorphic model of any complex system is, of course, a simplification of the actual system. Each reflects the level of discrimination, that is, the interest of the investigator.

The practical usefulness of homomorphic models can be assessed by comparing the informational needs of impact assessment with the needs of an individual going to a vast library. How would the probability of finding the necessary information by wandering randomly through library stacks compare with the probability of finding that same information by first using a library file index? Homomorphic models are the file indexes to existing information about complex systems. They are, in a sense, snapshots of an infinitely complex environment, each focusing on different scenes about which some knowledge is available.

Homomorphic models that are useful for describing our current and developing knowledge about the environment may be arranged in a simple hierarchy:

Physical environment
Abiotic component—nonliving, physical, and chemical things and processes
Biotic component—organisms and biological processes
Ecological component—living and nonliving things and processes that interact in a way that maintains a discernible dynamic system, for example, a wetland, a rainforest, and so on

Social environment
Personal component—specific individuals
Interpersonal component—interacting individuals and groups
Institutional component—assemblages of cultural rules of behavior that influence the dynamics of social systems, for example, political institutions, legal institutions, religious institutions, and so on

The subdivision of the environment into these six components does not imply that these subdivisions are essentially separate from one another. In fact, they are not separate, nor are they independent. This hierarchy, which is only one example of many that are possible, merely reflects the basic kinds of model available to help us understand various aspects of that complex, dynamically interactive system we call "the environment."

ITERATIVE ASSESSMENT PROCESS

Although many different definitions of environmental impact assessment exist, they share the same objective—providing insight to decision makers regarding the full range of consequences of their decisions. Historically, the process of impact assessment has been undertaken prior to and integrated with *project authorization* and has, therefore, focused on the decision making completed early in the project planning and design process.

In any proposed project that is likely to result in significant environmental consequences, myriad decisions must be made, not only in the early phases of project planning and design but also during property acquisition and management, construction, and operational and maintenance phases. **That all decisions are made and precisely known at the time of project authorization, and that they remain immutable from that point on, is extremely unlikely.** Any process of environmental impact assessment that does not take into account the often great differences between a planned project and that same project as it is actually implemented will, therefore, fail in its basic objective of informing actual decision making.

Rather than using environmental impact assessment solely as an adjunct to project authorization, developers can integrate it with project implementation as well as project development. This application requires a new perspective of impact assessment as an iterative process that retains its importance in project authorization (or denial) but also becomes a central tool for (1) monitoring and managing predicted impacts from the time of property acquisition and management through construction, operational, and maintenance phases; and (2) refining project development on discovery of impacts previously overlooked or changes in project design that may be required during actual implementation.

QUANTITATIVE AND QUALITATIVE ATTRIBUTES OF THE ENVIRONMENT

In the National Environmental Policy Act of 1969 (NEPA), the United States Congress clearly envisioned the assessment process as inclusive of quantitative analyses but certainly not exclusive of qualitative analyses. This intent is addressed most specifically in Section 102 of NEPA:

> All agencies of the Federal Government shall . . . utilize a systematic, **interdisciplinary approach** which will insure the integrated use of the **natural and social sciences and the environmental design arts** in planning and in decision making which may have an impact on man's environment; . . . identify and develop methods and procedures . . . which will **insure that presently unquantified environmental amenities and values may be given appropriate consideration** in decision making along with economic and technical considerations.

Despite these specific requirements, impact assessment during the first two decades of its development typically emphasized quantifiable physical, chemical, and biological entities and processes. When attention was given to the social environment, again, quantifiable attributes such as the various statistics of cost–benefit analysis or numbers of displaced households received primary if not sole attention. Typically nonquantifiable attributes or those not easily addressed by numerical analysis, including many sociological, political, and psychological factors, were ignored, except in some instances in which proposed projects were likely to have affected well-defined minority ethnic populations.

More recently, nonquantifiable considerations have become increasingly important, particularly in the Canadian approach to impact assessment, which places very heavy emphasis on public participation throughout the assessment process. Although public participation is typically required in impact assessment in other nations, the Canadian approach maximizes this participation and therefore gives excellent assurance that issues of concern to the public are given serious attention, whether or not those issues can be quantified.

Increased attention to social, psychological, and physiological well-being—the importance of which is often best evaluated not by arithmetic or statistical manipulations but by reliance on human values, morals, and ideals—has also been enhanced through the worldwide efforts of organizations such as the World Health Organization and the World Bank. In the United States, the growing concern regarding inequitable social distributions of environmental risks and benefits can also be expected to increase the emphasis given in impact assessment to environmental issues that are not easily or appropriately subject to numerical definition.

MITIGATION

Mitigation practices, including design, engineering, and management practices, are undertaken specifically to alleviate or enhance well-defined consequences of a proposed project. Although mitigation practices are ideally integrated into project design, they are by definition not intrinsically necessary for the achievement of the basic objectives of the proposed project; they are necessary, however, for the achievement of environmental objectives.

Because a specific mitigation practice is usually undertaken to meet a particular environmental objective (e.g., the design of a roadway underpass to minimize interference with migratory pathways of large mammals), impact assessors are tempted to evaluate that practice only in terms of its objective. However, mitigation practices may have many consequences that have little or nothing to do with their objectives and, sometimes, are even contrary to those objectives. For example, the roadway underpass may minimize interference of the roadway with the migration of large mammals; however, it may also serve as a device used by hunters to increase their kill of those same mammals. The conversion of a borrow pit to a wetland to enhance wetland habitat may result in an increase in waterfowl populations; it may also degrade the water quality of downstream water table wells. The use of rapid fertilization techniques may effectively minimize soil erosion on exposed banks; it may also result in undesirable runoff into receiving streams and aquifers.

SOURCES OF IMPACTS

Generic Types

Impacts may be described as *direct, indirect,* or *cumulative.* Direct impacts are changes in environmental components and processes that result immediately from a project-related activity or action. For example, the use of heavy equipment for clear-cutting (extensive cutting of trees) is likely to result in soil compaction; soil compaction is therefore a direct impact of clear-cutting. The same heavy equipment also results in temporary changes in the ambient noise level that are direct impacts of the use of that machinery. Clear-cutting itself, however performed, results in the removal of overstory (i.e., high vegetation such as tall shrubs or trees), a direct impact of the clear-cutting process.

Indirect impacts (sometimes called secondary impacts) are changes in environmental components and dynamics that are consequences of direct impacts. For example, soil compaction (a direct impact of the use of heavy equipment) may result in increased runoff and, thereby, increased risk for soil erosion. In this instance, increased runoff and soil erosion are consequences of soil compaction. Similarly, a change in ambient noise levels (a direct impact) may lead to changes in the migratory behavior of certain animals, such as birds, or in the disruption of schoolroom activities. These changes are indirect impacts caused by the change in ambient noise levels. The direct impacts of clear-cutting may result in a wide variety of indirect impacts, such as the loss of wildlife habitat, the introduction of light to the previously shaded forest floor, and changes in human land use.

Whereas direct impacts are environmental changes immediately linked to project activities or actions, **indirect impacts result from the varied interactions of direct impacts and the physical and social environmental components, processes, and conditions that are or become dynamically linked to those direct impacts.** In any impact assessment, indirect impacts are far more numerous than direct impacts and typically account for most of the assessment effort.

Each individual project results in direct and indirect impacts. Cumulative impacts are the aggregates of direct and indirect impacts resulting from two or more projects in the same area or region. For example, a highway project may result in the loss of 7% of the forest cover in a particular area, whereas a power line project undertaken 5 years later may result in an additional 3% reduction. The cumulative loss of forest cover over that period is, therefore, 10%. Assessment of cumulative impact is important because, whereas relatively small losses of forest land may not be environmentally significant on a per project basis, the cumulative loss may be highly significant.

Generally, environmental impact assessments have focused largely on those impacts that result from single projects. However, the growing concern for our global environment is likely to underscore the importance of assessing impacts in terms of whole developmental programs as well as individual projects. From a practical point of view, however, this goal is particularly difficult to achieve because of the different legal jurisdictions that exist among the different operational agencies of any national government, as well as because of different intra- and international interests.

Project Activities and Actions

Because direct impacts arise from project-related activities and actions, impact assessment must begin with the precise identification of those activities and actions. Moreover, as previously discussed, assessment of these activities and actions must encompass all phases of project development, including those associated with

- the planning, design, land acquisition and management, construction, operational, and maintenance phases;
- any in-place modifications of the final design or procedures for implementation; and
- any mitigation practices or procedures.

Some examples of activities and actions that might be associated with various phases of project development are included in (but are not limited by) Table 1.1. Specific activities and actions will, of course, vary with the type of project (e.g., transportation, energy) as well as with the type of environment surrounding the project (e.g., virgin rainforest, urban area).

Linking Impacts to Activities and Actions

The link between (1) the activities and actions that describe the project and (2) direct impacts on the environment lies in the precise manner in which those activities and actions alter the **material, energetic, informational, and/or perceptual conditions, attributes, and dynamics** of the physical and social environments. For example, direct impacts of dredging in a stream bed may, depending on the type of bed and the method used, include:

- increased concentration of suspended material (turbidity) in the water column, that is, a change in (a) the material attributes and

TABLE 1.1 Examples of Activities and Actions
Associated with Various Phases of Project Development

Pre-construction phases
   on-ground preliminary reconnaissance of site
   surveying of site
   collection of land-ownership records
   taking of test borings within proposed rights-of-way
   appraisal of real property
   negotiation with landowners
   relocation of displaced persons
   securing of physical access to site
   demolition and disposal of existing structures
Construction phase
   excavation
   pre-blasting
   dredging
   transport and placement of borrow
   clear-cutting and disposal of vegetation
   on-site transport and storage of materials and
     supplies
   dewatering
   on-site materials processing
   land clearing
   lighting
   structural fabrication and placement
   controlling runoff
   dust suppression
   landscape engineering and plantation
Post-construction phases
   selective cutting and disposal of vegetation
   application of preservatives
   application of pesticides
   disposal of runoff
   waste generation, storage, and disposal
   resource utilization
   process chemistry and other risk management
   management of by-products

    (b) the energetic attributes of the water column (due to the chemical energy available in organic muds)
- decreased organic substrate in the stream bed, that is, a change in (a) the material attributes and (b) the energetic attributes of the stream bed (due to the chemical energy available in organic materials)
- increased ambient sound levels (as a result of equipment operation), that is, a change in the energetic attributes of (a) the surrounding air

column (due to changes in soundwave amplitude and/or frequency)
and (b) the water column (due to vibrations)
- increased depth of the water column, that is, a change in the
  material attributes of the stream bed

Linking Impacts to Systems Interactions

The link between (a) direct and (b) indirect impacts lies in the dynamic
interactions among the various components of the physical and social environments. For example, an increase in turbidity in a stream (a direct impact
of dredging) may result in the abrasion of fish gills and, subsequently, a
decrease in affected fish populations. This effect would depend, of course,
on the species of fish, the abrasive index of the suspended particles, and
such factors as the velocity of the stream and the duration and timing of
the enhanced turbidity.

Should such turbidity-induced depression of a fish population occur,
and should that same species be of recreational or commercial value, additional indirect impacts of dredging might include a depression of recreational
or commercial fishing. Therefore, a depression of related economic conditions, with additional possible impacts on a wide range of personal, interpersonal, and institutional components of the relevant social environment,
might occur.

DIRECT AND INDIRECT IMPACTS: KEY CONSIDERATIONS

The distinction between direct and indirect impacts has absolutely
no meaning with respect to the actual environment. It is simply useful in
identifying how possible (i.e., plausible) pathways of causality may be initiated and promulgated. Figure 1.1 provides an example of a hypothetical
series of plausible pathways of causality, that is, a network of interrelated
causes and effects. Such pathways necessarily cut across disciplinary boundaries and force us to identify diverse factors and conditions that may make
a pathway more or less probable.

The objective of distinguishing between types of impact is not to
declare that one impact is direct and another indirect, but to organize our
analysis in a manner that insures that we will examine all possible effects of
proposed human actions on highly complex and dynamically interconnected
physical and social environments. The distinction between direct and indirect
effects is, therefore, simply a heuristic device for identifying the questions
that should be asked and the types of information that will have to be
collected and evaluated.

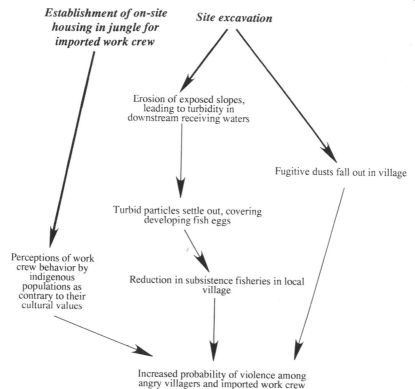

FIGURE 1.1 Network diagram depicting plausible lines of causality between project activities/
actions (italics) and physical and social environments. Boldfaced lines indicate direct impacts;
lightfaced lines indicate indirect impacts.

Whether direct or indirect, the word "impact" does not itself imply
any evaluation of the significance of that effect. An increase in turbidity in
a stream is not in and of itself positive or negative. A decrease in a biological
population is not in and of itself desirable or undesirable. These human
judgments must be imposed by a formal impact evaluation process; they
are not objective results of scientific or logical deduction. In short, impact
assessment should never be viewed as a means of avoiding making difficult
judgments. Instead, this process is a means of identifying the judgments that
must be made by decision makers.

OVERVIEW

The overview of the assessment process depicted in Figure 1.2 shows
that the process may be viewed as comprising four basic steps.

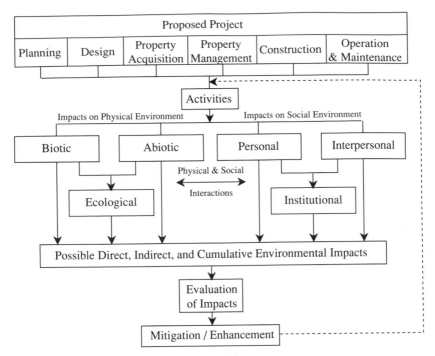

FIGURE 1.2 Overview of the assessment process.

1. Use our understanding of specific project-related actions and activities to identify specific direct impacts on the components and dynamics of physical and social environments. *Note:* The emphasis here is on establishing simple links between events that actually transpire during project development and the environment. This linking requires a precise understanding of how project activities and actions are actually carried out under local conditions (e.g., the machinery and procedures that will be used in this particular stream during the dredging operation).

2. Use our understanding of the dynamic interactions among physical and social components of the environment to identify subsequent indirect impacts. *Note:* The emphasis here is on specific dynamics that underlie the physical and social environment. This evaluation requires not only relevant disciplinary knowledge [e.g., the ecology of lacustrine (open water) impoundments] but also an interdisciplinary understanding of the local situation (e.g., economic and public health consequences of subsistence dependence on local lacustrine resources).

3. Apply a formal procedure to evaluate the possible environmental impacts we have identified (including direct, indirect, and cumulative impacts). *Note:* The emphasis here is on the identification of specific evaluation criteria, and of procedures for applying those criteria to establish the significance and value (desirability or nondesirability) of possible impacts. This most sensitive aspect of the assessment process requires careful consideration not only of local values, but also of regional, national, and even international goals and objectives.

4. On the basis of our evaluation of impacts, implement appropriate mitigation measures. *Note:* The emphasis here is on cost-effective engineering and/or management techniques to achieve stated development and environmental goals and objectives.

This overview focuses on the technical tasks of impact assessment and does not include various procedural aspects, such as the management of the interdisciplinary team, the use of public information programs, and the process referred to as "scoping." These and other procedural refinements are discussed in subsequent chapters, as appropriate. The first step of the assessment process, as depicted in Figure 1.2, is developed in greater detail in the remaining chapters of Part I. The second and third steps are developed for the various environmental components in Parts II and III. The final step is examined in terms of issues of special concern and is addressed in Part IV.

CHAPTER 2

ENVIRONMENTAL COMPARTMENTS AND DYNAMICS

When analyzing complex systems, we most often focus on highly simplified representations of those systems, otherwise called models. Environmental models are composed of interlocked cause–effect pathways. These pathways indicate the flow of materials, energy, or other forms of information from one component of the model to another, each component being a physical entity (e.g., a green plant, a city, a collection of molecules) or a property (e.g., buffering capacity of water, heat content of a solid). In considering any graphic representation of a model, a model component is thought of most usefully as a compartment in which entities or properties are contained.

As shown in Figure 2.1, a compartment of any model may be viewed as a singular entity or property with a defined input and output, or as a complex system composed of two or more interconnected parts. Each of these interconnected parts may, in turn, be composed of additional interconnected compartments and dynamics that are called submodels of the original simpler model. The more submodels considered, the more complex the basic model becomes, that is, the more variables and the more interactions among those variables which must be considered.

In environmental impact assessment, we must deal with perhaps the most complex system known in our daily experience; the integrated social and physical systems that surround us and constitute "the environment." However, **the objective of impact assessment is not to model the environment in all its infinitely complex detail.** Although increasing our precise understanding of the environment is of scientific interest, the practical goal in

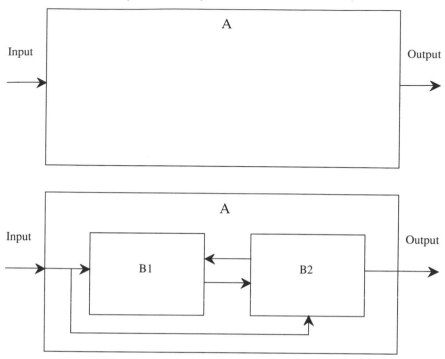

FIGURE 2.1   Example of a model of a system showing simple input–output model (*top*) and more complex model showing submodeling of transformation process (*bottom*). (*Top*) An input (energy or material) is transformed by component A into an output. No internal mechanism of A is considered. (*Bottom*) Same input, output, and transformation component (A), but the transformation is now submodeled to show complex interactions between B1 and B2.

impact assessment must be to apply the knowledge we already have to decision making.

How can we identify models of the environment that are most appropriate to a proposed project in a specific location? How can we quickly focus on those environmental pathways and variables that are of practical value for identifying the most likely direct and indirect impacts of that project? How can we deal reasonably with the multidisciplinary knowledge and jargon that must be consulted when exploring the interconnected physical and social environments?

PHYSICAL ENVIRONMENT: COMPARTMENTS AND DYNAMICS

Perhaps the simplest way to model the physical environment is to focus on (1) four compartments (geological, aquatic, atmospheric, and biological) and (2) the material and energetic flows and pathways that characterize each compartment and mediate their interactions. The distinction among the four compartments is, of course, artificial. After all, biological entities, air, and water occur within geological strata, just as biological entities, air, and soil particles occur in aquatic systems. However, the distinction is useful heuristically because it reflects the disciplinary compartmentalization of much of our knowledge and experience of the physical environment.

Any single compartment or combination of compartments of the physical environment contributes to or influences dynamic processes. Although such processes are identified and technically described in various diverse disciplinary paradigms, many may also be categorized and identified in nontechnical terms by their relation to (1) the movement of materials and energies through compartments and (2) the alteration of these materials and energies during their movement throughout the environment.

Whatever the factors that affect the movement and alterations of materials and energies in the physical environment may be, these factors exert their effects in at least six basic ways (Figure 2.2), by altering

- the **introduction** of materials and energies into environmental compartments,
- the **transformation** of materials and energies within environmental compartments,

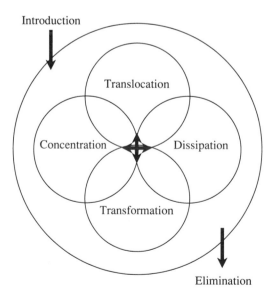

FIGURE 2.2 Generalized processes affecting the flow of matter and energy into and out of an environmental compartment.

- the **translocation** of materials and energies from compartment to compartment,
- the **concentration** of materials and energies within compartments,
- the **dissipation** of materials and energies within compartments, and
- the **elimination** of materials and energies from compartments.

Alterations of material and energetic introduction into environmental compartments may be manifest by changes in (1) types of materials and energies, (2) their concentration or amount, (3) the timing and duration of their introduction, (4) the place of introduction, and (5) the means of introduction.

Alterations of material and energetic transformations within compartments include changes in (1) types of transformations (e.g., physical, chemical, and/or biological), (2) rates of transformations, and (3) location of transformation within the compartment (e.g., upper atmosphere, bottom of water column, within the first few centimeters of the soil surface).

Factors that influence the translocation of materials and energies within a compartment are those that affect the movement of materials and energies without changing their physical, chemical, or biological characteristics. Thus, depending on the compartment, translocation may be affected by such factors as (1) temperature, (2) current velocity, (3) wind velocity, (4) animal migrations, and (5) opacity to sunlight.

Numerous physical, chemical, and biological factors influence the concentration or dissipation of materials and energies within a compartment. Regardless of the specific mechanisms by which materials and energies are concentrated or dissipated in compartments, project-mediated changes in these mechanisms may be causally related to such factors as (1) available surface area for absorption, (2) partition coefficients, (3) ion-exchange capacity, (4) dilution potential, and (5) oxidation–reduction potential.

Finally, materials and energies are eliminated or lost from one or more compartments through numerous mechanisms, including evaporation, physical conveyance (e.g., the flushing of sediments out of a stream during freshet flow), and the migration of organisms out of a particular region.

The six generic dynamic processes identified in Figure 2.2, as well as the various factors that may influence those processes, are not offered as fully comprehensive categories that completely define the complexity of the physical environment. However, they can be useful to the assessment team for identifying *disciplinary cognates*, that is, dynamic processes of particular disciplinary interest.

For example, *diagenesis* is a geological term denoting physical and chemical changes that can take place in sediments between the time of their *deposition* and their *solidification*. Some of those changes may be chemical transformations that may result in the introduction of chemicals into an overlying aquatic compartment. As a dynamic process, diagenesis can be influenced by various project-related activities, such as the clear-cutting of land and the subsequent release of soil particles into water.

Disciplinary Cognates of Generic Environmental Dynamics

Although the specific site of the proposed project and the specific nature of that project must determine which abiotic and biotic processes are relevant for a given impact assessment, the six generic dynamics and four environmental compartments discussed earlier should be considered.

Examples of typical abiotic and biotic mechanisms that accomplish the introduction of materials into soils, for example, include the *deposition* of organic materials, such as animal wastes and vegetative detritus accumulated during ecological succession; *precipitation* of inorganic and organic materials from the atmosphere; *weathering* of previously covered or sealed substrates; *settling* or dustfall of wind-dispersed fine materials; and impingement of *photic and thermal energies* from the sun.

Some of these same processes are involved in the introduction of materials into water (e.g., *deposition, precipitation*). However, surface waters also receive inputs through the *suspension* and *dissolution* of contaminating materials in tributary waters, and from the overturn (*destratification*) of impounded water that results in the introduction of nutrients from bottom muds (via diagenesis) into surface levels. In estuaries, tides are important inputs of energy and accomplish the work of fertilization (an introduction of materials) as well as waste removal (a dissipation and/or elimination process).

The atmosphere receives water (an introduction) through the process of *evapotranspiration,* which is the sum of water lost from the soil through *evaporation* and the *transpiration* (a translocation) of water through plants. The atmosphere also receives other materials because of *convective and advective transport* (translocations) of organic and inorganic materials.

The introduction of materials and energies into organisms is accomplished by such mechanisms as *feeding, respiration, osmosis, absorption through the skin,* and *auditory, visual, and tactile stimulation.*

Once materials have been introduced into soil, water, air, and organisms in the project area, they are variously transformed, translocated, concentrated, dissipated, and eliminated, depending on a variety of individual and interrelated abiotic and biotic factors.

For example, detritus (dead organic material) that is deposited on surface soils by overhanging vegetation may be transformed or altered by *biological decomposition, chemical oxidation, physical disintegration,* and *leaching.* Leaching (of dissolved materials) and *eluviation* (of suspended materials), and subsequent *percolation* of surface water that contains both dissolved and suspended materials, also accomplish the translocation of materials through different soil horizons. As a result, these processes cause the introduction of such materials into groundwater. Translocation of surface-deposited materials may also occur through *erosion*.

Sequestration (or entrapment) of materials by soils may be accomplished through *adsorption* onto particle surfaces, *absorption* into soil particles, *chemical chelation,* and other mechanisms. Chemical chelation also occurs in water, as does physical sequestration by means of *sedimentation*. Materials may be sequestered in individual organisms by the process of *biomagnification,* and by *metabolic transformations* accompanying biological productivity. These materials may be eliminated from plants through *transpiration* and *guttation* and from animals through *excretion*.

These and other examples of specific dynamic processes in the various environmental compartments are included in Table 2.1. The importance of constructing a list (such as Table 2.1) of ongoing abiotic and biotic processes in any proposed project area cannot be overemphasized. Such a list provides the basis for

- identifying key dynamics considered important by various specialists who have disciplinary and/or legal jurisdictional interest in specific environmental components and phenomena,
- identifying (by consulting specialists and the relevant literature) those variables and factors that are generally considered important for understanding specific dynamic processes,
- correlating specific project-related actions and activities to such variables and factors, and
- defining plausible mechanisms by which project effects on a dynamic process in one environmental compartment can affect other dynamic processes in the same or another compartment.

Linking Projects and Physical Impacts

Potential impacts of project development on the physical environment may be identified by first identifying how different project actions or activities may influence dynamic processes in the soil, water, air, and organisms in the project area. *Probable impacts* can be estimated only in terms of the data that are specific to the proposed project and to the project site. Therefore, potential impacts are those impacts that are reasonable in light of theory or general understanding; probable impacts are those that are reasonable in light of theory and site-specific conditions. The identification of potential and probable impacts on the physical environment depends on an informed understanding of specific dynamic processes.

TABLE 2.1 Examples of Dynamic Processes That Characterize Four Basic Environmental Compartments

Generic dynamic processes	Environmental compartments			
	Geological	Aquatic	Atmospheric	Biological
Introduction	Illuviation Precipitation Burrowing Percolation	Destratification Precipitation/deposition Tidal flow Entrainment	Advection Convection Fugitive emission Evaporation	Feeding/respiration Osmosis Absorption Sensation
Translocation	Infiltration Erosion Leaching Eluviation	Diffusion Hydraulic flow Turbulence	Inversion Advection Convection	Transpiration Diffusion Circulation
Transformation	Oxidation Diagenesis Weathering Compaction	Oxidation/reduction Diagenesis Mineralization Metabolism	Photo-oxidation Catalytic oxidation	Photosynthesis Respiration Metabolism
Concentration	Chelation Adsorption Channeling Ion-exchange	Chelation Sedimentation Stratification	Stratification Interception Convection	Biomagnification Absorption
Dissipation	Evaporation Leaching Eluviation Erosion	Biomagnification Dilution Absorption Adsorption	Dilution	Excretion Guttation Detoxification Catabolism
Elimination	Evaporation Leaching Eluviation Erosion	Evapotranspiration Hydraulic flow Drawdown	Fallout Condensation Precipitation	Excretion Guttation Transpiration

For example, turbid particles are introduced naturally into a stream through several biotic and abiotic processes. Construction activity such as clear-cutting, excavation, dredging, and landscaping may also introduce such particles. Project-induced turbidity may be different from natural turbidity in terms of type (i.e., organic or inorganic), amount, size, and other physical characteristics of the particles, and timing of introduction. Regardless of source, turbid particles in water will be translocated, transformed, concentrated, and/or dissipated according to any number of specific and interconnected biotic or abiotic dynamics.

Particles may, for example, settle out (sediment) rapidly (e.g., under low hydraulic gradient and high specific gravity). Consequent siltation or downstream aggradation may affect other processes, including the diagenesis of downstream muds. One of the possible consequences of the diagenesis of sediment is the release of plant nutrients from that sediment to overlying waters. Thus, if project-induced sediment is organic, mineralization (or decomposition) of that sediment may lead to an increase in downstream concentrations of plant nutrients. A biological consequence of enhanced nutrient levels may be an increase in the production of plant biomass. Whether or not such a sequence of events will actually occur (or be detectable) depends on various factors, including ambient concentrations of nutrients, type of aquatic vegetation, and amount, type, and timing of nutrient release through diagenesis.

Other potential impacts of introducing turbid particles into surface water include abrasion and/or coating of fish gills, with consequent impairment of the respiration of fish; covering of developing fish spawn, with consequent suffocation of developing fry; alteration of downstream benthic (bottom) habitat and, thus, of the benthic community; and reduction of sunlight penetration through the water column and, thus, a change in thermal regime as well as a reduction in the photosynthetic production of oxygen at lower depths. Again, whether or not any of these potential impacts on abiotic and biotic aspects of the aquatic environment are probable impacts depends on specific data. Examples of relevant data include particle size, existing fish species, timing and nature of fish reproductive behavior, abrasive and chemical characteristics of turbid particles, existing downstream benthic habitat, and ambient turbidity and its relationship to the existing photic and thermal regimes of receiving waters.

SOCIAL ENVIRONMENT: COMPARTMENTS AND DYNAMICS

Despite the almost quarter century of experience in formalized environmental assessment procedures, the focus of the assessment effort has remained on the physical environment. Direct, indirect, and cumulative impacts on the social environment typically have been recognized and evaluated (if at all) within the relatively narrow constraints of economics, aesthetics, and archaeology. However, growing global concern about public health issues, environmental equity, cultural pluralism, and environmental sustainability underscores the necessity for overcoming our heretofore limited sociological perspective.

Although the diversity and complexity of social reality apparently has no limit, this fact does not preclude an approach to assessment that is parallel to the approach taken in assessing impacts on the physical environment, that is, using highly simplified, heuristic models to identify basic compartments and dynamics, using these models to identify disciplinary cognates, and, finally, using our disciplinary knowledge (however limited) as well as our personal experience to identify and evaluate plausible direct and indirect links between proposed projects and social systems.

As shown in Figure 2.3, relationships between the physical environment in which people live and the behavioral rules and ideological values with which people are enculturated constitute their *cultural base*. How that cultural base is actually manifested in an orderly and patterned scheme of human activity reflects the *social organization* of a people. Although the ways in which a general cultural base can be translated into specific patterns of social organization appear to be infinite, we can assume that these ways all derive from a variety of functional needs, including (1) maintenance of health and well-being, (2) division and coordination of labor, (3) communication, (4) training of new members, (5) regulation of conflicts, and (6) distribution of power and wealth.

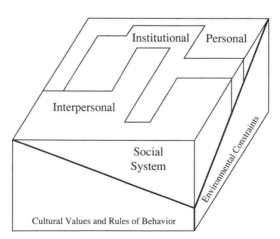

FIGURE 2.3   Overview of the integrated components of a social system.

For purposes of impact assessment, we must consider that projects exert their impacts directly or indirectly on specific personal, interpersonal, and institutional components of social organization, as well as on their interactions (Figure 2.4). These impacts may be described in terms of their influence on the attributes, conditions, and dynamics of each component.

Attributes, conditions, and dynamics appropriately considered at the personal level include a wide range of emotions, interests, values, beliefs, attitudes, and experiences. Specific examples are fear, anxiety, and anger; the sense of self-esteem or worth; individual lifestyle; health and safety; housing; personal well-being and security; the sense of fulfillment or promise; the personal investments and perquisites of self-image and role; and the personal sense of right and wrong and of good and bad.

At the interpersonal level, a level at which the focus is not on the single person but on small and large groups of persons, conditions and dynamics appropriately considered include community identity; social distributions of opportunities, risks, rewards, and authority; intergroup conflict and resolution; the legitimacy of power; opportunities for the generation and enjoyment of wealth; technological development; access to communication, transportation, housing, and health care; the definition of public objectives; educational and vocational opportunity; public safety; rights and duties; and public morality.

Finally, at the institutional level, impacts may be expressed in terms of the diverse means, mechanisms, practices, and procedures integral to the definition and promulgation of political, economic, religious, and legal objectives.

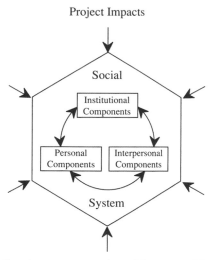

FIGURE 2.4 Projects affect the components of a social system and their dynamic interactions.

Examples of Social Impacts

Figure 2.5 depicts three alternative locations of a hypothetical project with respect to three individuals. As depicted in this figure, individuals B and C are geographically equidistant from individual A. However, in terms of social reality (i.e., reality as perceived by these individuals), the *social distance* between individuals A and C (as measured by frequency of visitation, importance of visual contact, sharing of values and beliefs, etc.) is depicted as much shorter than the social distance between individuals A and B.

Because of the different social distances between these individuals, different project locations can have very different social impacts. As shown in Figure 2.5, alternative location 2 may directly interfere with interactions between individuals A and C. Such interference may occur as a result of interrupting the visual line-of-sight between them, or by making it more difficult and time-consuming for each to visit the other. This interference could also result, at the personal level, in increased anxiety or even anger in the affected individuals. On the other hand, alternative location 1 may have no effect on the social distance between individuals A and B. Alternative location 3 may similarly have no effect on the social distance between individuals A and B, and may also minimize (although not remove) project obstruction of visual or personal communication between individuals A and C.

Of course, restricting consideration of project impacts on social distance to impacts on the social distance between individuals is not warranted.

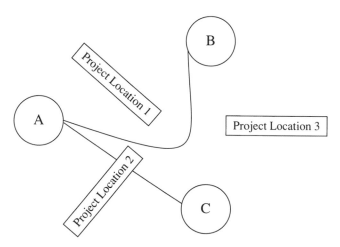

FIGURE 2.5   Alternative locations of a project that have different effects on the social distance between individuals (A, B, and C).

The degree and frequency of an individual's participation in or identification with groups and organizations in the general project can also be affected by project development and operation. The important point is that the social distance between any two or more social components (whether individuals or groups) can be altered by project location.

Projects may also, of course, effectively shorten social distance between social components. The shortening of social distance may (1) facilitate the communication between individuals and groups who desire improved access to one another and (2) facilitate (or even force) communication between individuals and groups who do not desire improved access to one another. For example, a project may result in the removal of a natural physical barrier (e.g., a stand of trees) between people who, except for that barrier, might enjoy visiting one another or might be pleased by the ability to see anyone else, simply to feel less isolated. However, the same project, with the removal of the same natural barrier, might result in clashes between groups whose disparate interests and life-styles were previously shielded from one another.

The concept of social distance is, of course, only one of many concepts that are useful for describing dynamic interactions among individuals or groups. Another important concept is the concept of *role* (i.e., the expectations that others have of our behavior and which expectations we internalize). Consider, for example, that a proposed project will require a large amount of construction work in an isolated but populated area. This construction work (e.g., associated with the construction of a major pipeline or oil field) may result in a significant change in the distribution of money in the local area. The income of the individuals who do the work (likely to be imported specialized labor) and the local people who service them (food, housing, recreation, etc.) will probably be very high compared with that of other local people. However, because of the general infusion of monies, the cost of living for all members of the community may increase dramatically, with a consequent severe reduction in real income for individuals not directly or indirectly involved with the construction work. Those individuals who perceive their primary role within the family as that of "breadwinner" might experience serious personal (psychological) and interpersonal (familial) dysfunctions as a result of such a serious drop in real income.

Linking Projects and Social Impacts

As for the physical environment, projects result in direct and indirect impacts on the social environment. Moreover, direct impacts on social components and dynamics can lead to important indirect impacts on the physical environment, and direct impacts on physical components and dynamics can lead to important indirect impacts on the social environment.

For example, the soil deposition of airborne anthropogenic contaminants (e.g., nitrous and sulfurous oxides produced by a petrochemical power plant) may result in diminished long-term capacity of those soils to support vegetation, because of acid-induced leaching of plant nutrients. The human use of these increasingly nutrient-depleted soils can, therefore, be expected to change, with a wide variety of personal, interpersonal, and institutional ramifications, depending on the specific cultural or social milieu of the affected peoples.

Similarly, the displacement of individuals (even small numbers) as a direct or indirect consequence of project development (e.g., property sale or condemnation; increased land values) may result in various forms of *social disorganization,* depending on the roles such affected individuals play in the social dynamics that characterize the primary and secondary group structures of the community (Figure 2.6). In turn, behavioral cognates of those changes in attitudes, values, and beliefs that reflect social disorganization may have pronounced effects on the physical environment, including overutilization of previously conserved resources and changes in land use.

A particularly important type of social impact is exemplified by *environmental inequity,* that is, the unequal social distribution of environmental

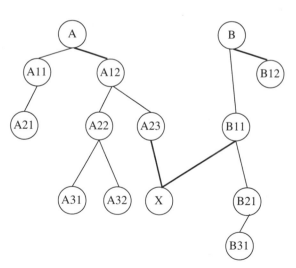

FIGURE 2.6   Example of a communication network that is useful for depicting interpersonal interactions. Thickness of lines connecting individuals may indicate frequency of communication or some other measure of interaction. Whereas individuals A and B have no direct connection, they are nonetheless part of the network because of their indirect relationship with individual X. If individual X is deleted, two distinct and noninteracting groups emerge.

hazards, risks and benefits. The issue of environmental equity encompasses many different social and political concerns, including

1. **Economic/racial inequity:** The siting of hazardous chemical disposal facilities primarily in economically depressed areas, resulting in significantly elevated risks of chemical exposure among minority and politically unempowered populations.
2. **Cultural inequity:** The uncontrolled development of rainforests or other resources, resulting in the dislocation of aboriginal populations dependent on those resources.
3. **Sexual inequity:** The reliance on current health standards regarding chemical exposure of both males and females, which are preponderantly based on workplace experience with only males.
4. **Age or generational inequity:** The use of technologies that result in wastes that cannot be adequately treated but must await future technology, resulting in the buildup of environmental risks and hazards for future generations.

Although such environmental inequities may be physically manifested in terms of the physiological health of affected populations, they are not relevant only to human health. These inequities are also relevant to human perceptions, values, and beliefs and are, therefore, of important psychological, sociological, and political consequence.

CHAPTER 3

METHODOLOGIES FOR ASSESSING IMPACTS

Whatever the particular legal basis of environmental impact assessment, the assessment of impacts always requires a wide range of diverse efforts that must be directed toward the timely achievement of the following objectives:

- the identification possible impacts,
- the evaluation of identified impacts, and
- the mitigation of significant impacts.

The various methods, principles, and rules for achieving these objectives may be usefully described as *managerial, analytical,* and *integrative efforts.*

MANAGERIAL EFFORTS

Environmental impact assessment requires inputs from and cooperation between a number of different specialists. Therefore, the temptation exists to define the requirements of impact assessment in terms of the requirements (data, time, budget, etc.) of each individual specialty. Within the real constraints of time, money, and personnel, such a perspective is likely to lead to serious and disruptive confrontations between various contending interests.

An alternative approach involves considering that impact assessment minimally requires the cooperation and coordination of individuals toward a common goal, that is, informing decision makers of the full consequences of their actions. Such an approach to defining the assessment process, which emphasizes the basic task of getting things done through people (and not specialties), highlights the assessment process as a *management intensive process.*

The manager of any project plans, organizes, staffs, directs, and controls the activities of others. Thus, the manager of the impact assessment process is ultimately responsible for the efficiency, the relevance, and the adequacy of the assessment process.

Each analytical and integrative effort of the assessment process may be subdivided into specific tasks and subtasks. For example, specific tasks performed as part of the overall effort to collect environmental data may include the collection and compilation of data and information on such individual environment components as

- surficial and subsurface hydrology
- soil types
- regional flora and fauna
- macro- and microclimatology
- special or unique habitats
- demographic patterns and projections
- recreational uses of resources
- water quality and quantity

The performance of each task requires personnel, time, and money. In making decisions that affect the allocation of personnel, time, and money, the manager of an impact assessment must continually balance two considerations. First, what are the total available budget and personnel resources available to the assessment effort? Second, what does each expenditure of money and effort specifically buy in terms of useful input into the decision-making processes of project development? Allowing individual specialists to go off on data-collection sprees is easy. Organizing and controlling data-collection tasks to insure that, on their completion, a sufficient budget remains to undertake those other analytical and integrative tasks that must be performed in the process of impact assessment is difficult.

Although no single best way of managing an environmental impact assessment has been determined, certain attributes of a well-managed assessment may be identified and summarized.

1. Individual tasks are identified and described early in the assessment process. The description of each task includes specifically what is to be done, precisely why and how it is to be done, and by whom. The task description also specifies the product or "deliverable" that will be in hand on the completion of the task (e.g., a report, a map, a graph, a file index). Finally, the task description includes the expected number of person-hours or -days for task completion, the time frame over which those person-hours are to be extended, and the link (if any) between that task and other tasks undertaken concurrently or subsequently. Only by such a careful definition of tasks will the assessment effort become a truly interdisciplinary effort

rather than an uncoordinated, multidisciplinary exercise of no practical relevance to decision making.

2. Before any specific task is actually undertaken, a summary is made and compiled of person-hours and time frames required for all analytical and integrative tasks. This summary will allow the interdisciplinary team to visualize the overall assessment project, and to make the revisions and adjustments considered necessary to insure an efficient, comprehensive, and useful impact assessment. Once refined, the summary can be used to compile a schedule for the actual assessment process.

3. The schedule for the overall assessment process will also specify person-hours and time frames for in-house team conferences, internal and external liaisons, and any public meetings and hearings.

Task summaries and schedules may be depicted in a number of ways. One example is provided in Figure 3.1. Some graphic representation of these tasks must be made available to the whole team so each individual can easily identify (1) his or her individual responsibilities and (2) how his or

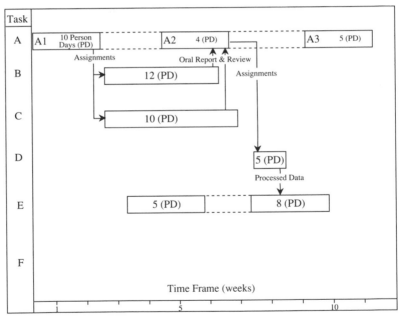

FIGURE 3.1 Graphical representation of assessment task schedules. Task A may be, for example, the management of the assessment effort; other tasks may include data collection for physical and social resources, assessment of public health impacts, and all activities related to liaison with governmental agencies and the public. Bars show ongoing assessment effort; broken lines indicate suspension of effort. The scheduling of specific work assignments and reports is indicated by arrows.

her responsibilities and those of other team members are interrelated and interdependent.

Of course, impact assessments must deal with many unknowns. Therefore, one may question the importance of such a careful consideration of schedules early in the assessment project. The reason is obvious to anyone who has had experience with any effort constrained by time and money and by the necessity of providing a product that is directly useful to decision makers. **Because we cannot foresee everything that might happen, we must schedule what we do know will happen. Otherwise, we cannot reserve sufficient budget, time, or personnel to deal adequately with contingencies.**

The good manager of an assessment assumes that something will go wrong; that required data will not be as easily available as first assumed; that certain real impacts will be completely overlooked and will require careful consideration; and that differences in personalities, values, and styles among team members and others utilized for assessment purposes can cause serious problems and delays. These events can and will happen, regardless of whether carefully planned schedules or carefully defined tasks have been prepared. However, without such a schedule, these unplanned events are more likely to happen, and are more likely to happen when the probability that time, money, and personnel have already been committed is greater.

In many instances, impact assessment is conducted, completely or in part, by contractors or individual consultants to a governmental agency that has assessment responsibility. In such instances, agency personnel typically defer to the contractor or consultant with respect to the day-to-day activities of assessment, which is often a serious mistake. An agency-based authority is best advised to participate in and exercise oversight authority over the detailed definition of assessment tasks, schedules, and products.

ANALYTICAL AND INTEGRATIVE EFFORTS

Analytical efforts are those directed toward the description of component parts of systems. Integrative efforts are those directed toward the holistic description of systems. Environmental impact assessment requires both types of effort.

The following questions are typical of analytical questions that may be addressed when assessing impacts, for example, on surface waters.

- What is the ambient summertime concentration of nitrate nitrogen within the first several meters from the surface?
- How does turbidity vary with surface runoff?
- What microscopic and macroscopic species utilize which aquatic habitat?

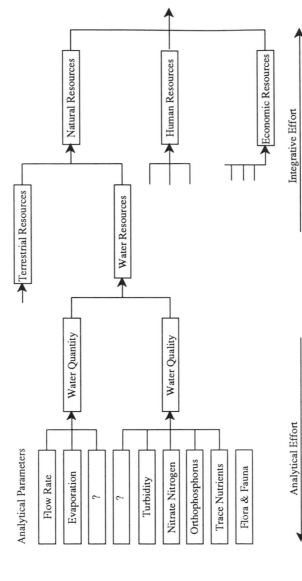

FIGURE 3.2 Analytical and integrative effort required by the assessment process. To insure the practical utility of assessment, an appropriate balance between analytical and integrative efforts is vital.

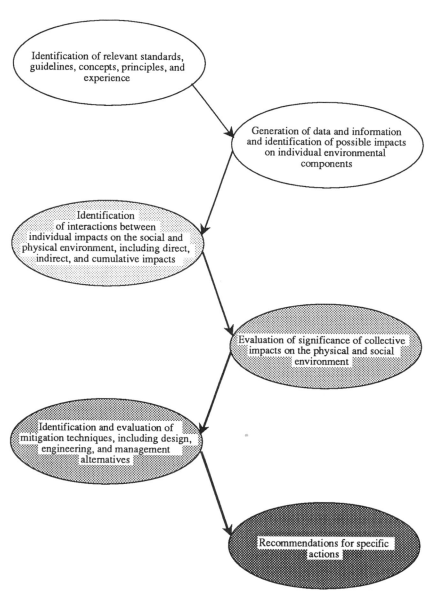

FIGURE 3.3 Linked analytical and integrative tasks. Enhanced shading indicates relevance to the objective of the assessment process.

As depicted in Figure 3.2, many physical and chemical parameters may be analyzed in any aquatic system. However, analytical data, such as 0.15 ± 0.4 mg/liter NO_3-N (i.e., a range of nitrate nitrogen concentrations), do not inform decision making, which is the objective of impact assessment. The analytical data must be processed—selected and integrated so conclusions can be made with respect to water quality. For example, is the nitrate nitrogen concentration within the standards for drinking water? Does the concentration promote high productivity in the water column? Does the concentration indicate agricultural or industrial runoff? Other analytical data, such as flow rate, evaporation rate, and downstream discharge, must be similarly selected and integrated so conclusions can be made regarding water quantity. The subsequent integration of judgments made regarding water quality and quantity allow judgments to be made about water resources. As shown in Figure 3.2, still higher levels of integration are necessary for decision making to be informed by a comprehensive appreciation of the total environment.

The assessment process may be viewed as a series of linked tasks (Figure 3.3), beginning with highly analytical tasks and progressing through increasingly integrative tasks. Analytical tasks are conducted at the disciplinary or multidisciplinary level, that is, specialists focus on their relatively narrow areas of interest and expertise. Here the emphasis is on specific parameters and on dynamic processes in which those parameters take part. Integrative tasks are conducted at the interdisciplinary level. Here the focus is on the interconnectedness of diverse environmental components. Emphasis is on the dynamic linkages between disparate parameters and dynamic processes that transcend any single discipline.

METHODOLOGIES

The methodologies that are clearly associated with impact assessment, although they are not unique to impact assessment, are of four types:

1. **Overlay methodologies:** These require physical or computerized overlays of individual maps of social and physical attributes of the project area (e.g., soil types, demographics, geological features, surficial hydrology)
2. **Checklist methodologies:** These itemize environmental parameters to be investigated for possible impacts
3. **Matrix methodologies:** These correlate cause–effect relationships between specific project activities and impacts
4. **Network methodologies:** These define a network or "causal chain" of possible impacts that may be triggered by project activi-

ties and that require the analyst to trace out project actions and direct and indirect consequences

Within each of these categories are numerous variants, each adapted for specific purposes. Although most of these methodologies or techniques have been used primarily to identify possible impacts, all can be (or have been) adapted for evaluating impacts and communicating the findings of impact assessment to the public and to decision makers.

Overlay Methodologies

Overlay techniques require a well-developed database that is amenable to graphic representation on maps of the project area. Typical data include:

- topological data
- land and resource use patterns
- surface and groundwater aquifers
- habitat types
- air dispersal patterns
- demographic data
- historic and archeologically important sites
- remote sensing data, for example, color-infrared imaging of vegetative productivity, soil moisture content

Coupled with currently available computer technology, which allows for a wide range of graphic presentations, this technique is particularly effective in identifying possible interrelationships of diverse components of the total environment. For example, project locations immediately upstream of surface and groundwater aquifers would obviously highlight the issue of aquifer contamination by project-related runoff. Visual data on habitat types, land use, and other information derived from remote sensing can be combined to identify areas to be given special attention during the progression of field studies.

Because visualization can easily prompt questions related to the interrelationships of otherwise disparate components and attributes of the environment, this technique has broad utility in the analytical and the integrative phases of impact assessment. Of particular importance is its potential for (1) identifying geographic areas and environmental issues of particular concern early in the assessment process (see "Scoping Process," Chapter 6) and (2) communicating the result of the assessment process to the public and to decision makers.

The method is limited, of course, by the availability of diverse databases and of the appropriate computer hardware and software. Note also that, as useful as this technique is for suggesting possible broad impacts that might reasonably be considered, it does not perform any analysis of possible impacts or evaluate those impacts.

Checklist Methodologies

Ranging from very simple to very complex, checklist techniques essentially depend on a list of environmental components, attributes, and processes that are subsumed under more-or-less disciplinary headings, such as Geology, Vegetation, Animals, Water, Air, Services, and so on. The list, therefore, serves as a reminder to the assessment team of "things to consider."

In some instances, the components, attributes, or processes under each heading are cross-referenced to procedures, factors, and issues that are relevant to the appropriate assessment. These lists are often referred to as *descriptive checklists*. In other instances, the checklist is presented in the form of questions that are typically answered "Yes," "No," or "Unknown." Although such *questionnaire checklists* require the analyst to consider specific subject matter, they do not give direction on how to analyze that subject matter with respect to a given project.

Checklists to be used for the identification of possible impacts are most often composed of items or questions that are generally applicable to proposed projects, that is, most checklists are not project specific or geography specific. However, checklists designed for specific projects and geographical areas will become more readily available as impact assessment is increasingly used as an aid in decision making.

Whereas the vast majority of checklists is used as an aid in the identification of impacts, a few have been specifically designed to conduct an evaluation of impacts. For example, the United States Federal Highway Administration (USFHWA) has developed a questionnaire checklist that uses simple responses ("yes," "no," "unknown") to evaluate wetland functions before and after proposed project development. This method is available in text and computerized formats. Similarly, the United States Fish and Wildlife Service (USFWS) and the United States Army Corps of Engineers (USACE) have devised questionnaires useful for the evaluation of wildlife habitats. These questionnaires are known as the Habitat Evaluation Program (HEP; USFWS) and the Habitat Evaluation System (HES; USACE). Numerous other questionnaires have been developed by academic authors and governmental agencies.

The primary limitation of checklist methodologies is that they do not correlate particular types of impact with specific activities related to the various phases of project development. Also, as static lists, they tend to promote the notion that all possible environmental impacts are already known and that the task of the assessor is simply to select from known impacts. Finally, because impacts are basically catalogued by disciplinary headings, checklists are not particularly useful for identifying possible indirect impacts that may result from systemic changes brought about by project development.

Matrix Methodologies

Matrices used in impact assessment are essentially modifications of checklists, that is, in addition to a vertical listing of impact typologies (e.g., increase in runoff, modification of nutrient regime) organized under componential headings (Water, Air, etc.), matrices contain a horizontal listing of project activities, ranging from early planning through operational and maintenance phases of project development. This approach facilitates relating specific project activities (clear-cutting, dredging, right-of-way management, etc.) to specific types of impact.

Matrices are sometimes used not only for the identification of possible impacts but also for evaluating those impacts, often by entering numbers into the matrices that represent subjective estimates of the significance of the relevant impacts. Summary addition of such numbers in matrices, completed for different design alternatives, has been used to compare the environmental significance of those alternatives.

Although matrices do associate project activities with possible impacts, they do not identify specific causal pathways through which impacts actually occur. Like checklists, matrices are not particularly useful for identifying indirect (or systemic) impacts. They also imply that all possible impacts are already known, regardless of the type of project or site-specific data and information.

Network Methodologies

As "causal chains," networks are extremely useful for identifying direct and indirect impacts. The network may begin with a specific project activity (e.g., use of heavy equipment during excavation) and trace through various direct impacts (e.g., increase in sound level, compaction of soils). The same network may subsequently trace secondary impacts from the various direct impacts (e.g., interruption of nesting behavior of migratory birds using habitat in the project area, decrease in soil percolation rates because of soil compaction). If the causal chains are pursued diligently, secondary impacts are seen to lead to tertiary impacts and on to higher order impacts.

A major advantage to network analysis is that it cleanly cuts across disciplinary lines. For example, if heavy equipment can lead to soil compaction (a direct impact), and if soil compaction can lead to a decrease in soil percolation (a secondary impact), which may result in an increase in surface runoff (a tertiary impact), subsequent higher order impacts may include changes in fisheries in waters that receive that runoff and subsequent economic impacts of changes in those fisheries. In this example, we begin with a consideration of soil but are subsequently directed to consider water, fisheries, and economies.

A second important advantage of network analysis is that it forces the identification of site-specific factors and conditions necessary for the establishment of a proposed cause–effect relationship. For example, if increased runoff into a receiving stream is proposed as a possible cause of a change in fisheries in that stream, certain data and information—such as the species of fish in that stream; their requirements with respect to oxygen, temperature, and habitat; their particular sensitivities; and the factors related to increase runoff (the amount and nature of suspended particles, flow rate, chemical constituents, etc.) that may or may not interfere with their requirements or sensitivities—must be considered.

A third advantage is that, unlike checklists and matrices, this technique does not imply that all possible impacts are known. On the contrary, this technique depends absolutely on site-specific conditions and attributes and, therefore, offers a greater probability that the assessment process will be directly relevant to the proposed project.

A disadvantage of this technique is that the analyst must have a general knowledge of the various types of environmental components and dynamics. As discussed in Chapter 2, the premise of this volume is that this limitation can be overcome by focusing on certain generic dynamics that can be applied to the various compartments of the physical and social environment. This theme will be developed more fully with respect to network analysis in Parts II and III. A second disadvantage, at least as perceived by many, is that network analysis does not easily lend itself to the evaluation of impacts. Nothing inherent in network analysis gives more or less significance or value to one "causal chain" over another.

SOME ADDITIONAL CONSIDERATIONS

As mentioned previously, these techniques are not unique to environmental impact assessment. These methods are, after all, information handling and management techniques useful for assessing changes in complex systems. As such, they are also useful for other types of assessment that have become, since the advent of environmental impact assessment, increasingly important: environmental health impact assessment, hazard assessment, and risk assessment.

Environmental health impact assessment focuses on **human health and safety risks that may be associated with environmental changes resulting from project development.** Hazard assessment is the effort to identify **possible sources of physiological, physical, and psychological harm or injury.** Risk assessment seeks to assign **probabilities to the occurrence of harm or injury to selected populations.**

One measure of the historical inadequacy of environmental impact assessments is that the World Health Organization has found it necessary to emphasize the importance of human health and safety considerations in environmental impact assessment by calling special attention to environmental health impact assessment as a necessary activity in and of itself. Although the failure of many impact assessments to consider human health may be attributed largely to a preoccupation with identifying and evaluating impacts on the physical environment, the fact remains that, conceptually, environmental impact assessment cannot give the physical environment any priority over the human environment, particularly those issues of human health and safety. For the purposes of this volume, the human health dimensions of environmental impact assessment, which are as amenable to assessment by means of the methodologies and techniques just discussed, are discussed in Chapter 13.

Hazard and risk assessments have become major concerns because of our growing awareness of chemical and other potential hazards that might be associated with electromagnetic radiation and food chain contamination. Again, the methodologies typically employed for purposes of environmental impact assessment can and should be applied to meet the objectives of hazard and risk assessment. In fact, in some countries (e.g., Malaysia), hazard and risk assessment are required by law to be included in environmental impact assessment. In others (e.g., the United States), hazard and risk assessment are included as a matter of departmental policy (e.g., Federal Highway Administration). These topics are discussed in Chapter 17.

CHAPTER 4

ENVIRONMENTAL STANDARDS

To identify the important uses and misuses of environmental standards in the conduct of environmental impact assessment, differentiating *standards* from *parameters* and *criteria* is instructive.

Environmental parameters are attributes of the environment that may be measured quantitatively or defined qualitatively. Concentrations of chemicals (e.g., dissolved oxygen in water, nitrate in soil, sulfate in air, organic mercury in living tissue), physical attributes (e.g., percolation rate in soils, shoreline development in impoundments, temperature, photoperiod), and biological parameters such as population density and birth and mortality rates are all parameters that may be quantitatively measured.

Qualitative parameters include environmental attributes such as aesthetic quality, community (human) cohesion, ecological health or well-being, and anxiety. Of course, numerous procedures have been developed to quantify various qualitative parameters. The basic difference between quantitative and qualitative parameters is that the former are amenable to quantitative analysis by means of accepted standard methods and procedures, whereas the latter, if quantified at all, are quantified by methods and procedures that are not universally accepted.

Whether quantitative or qualitative, the number of environmental parameters is essentially infinite, dependent only on our inventiveness.

Environmental criteria (Figure 4.1) constitute a subset of environmental parameters, that is, criteria are selected parameters. These features are selected because they are generally accepted as important for understanding phenomena that are of particular interest to humans. In this sense, criteria are often described as parameters to which social objectives have been appended. Criteria also typically include a judgment about a desirable limit

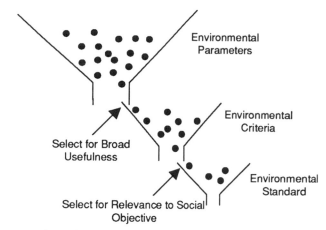

FIGURE 4.1   Relationships among environmental parameters, criteria, and standards.

(expressed quantitatively) or condition (expressed qualitatively) that is intended to promote a specific use of an environmental resource.

For example, the criterion for the pH of water used for swimming may be expressed as ≥ 6.8 and ≤ 7.2. The purpose of this criterion is to prevent the occurrence of "swimmer's eye," a condition caused by a pH-mediated chemical reaction on the surface of the eye. A criterion for pH in water supplies used for other purposes would, of course, express a different pH requirement.

Although criteria represent a technical consensus on desirable attributes of resources and their potential uses, criteria generally do not denote any legally enforceable limits. They are typically only advisory.

Environmental standards, however, are precisely those criteria that have been adopted by legal authority. In this sense, then, standards constitute a subset of criteria that have been adopted or modified with the intent of legal enforcement.

SOME PRACTICAL CONSIDERATIONS

Given the relationships among environmental parameters, criteria, and standards, various problematic issues arise regarding the use of standards for purposes of impact assessment. These issues are of several kinds, including the changeability of standards, the compatability of standards, and the scientific basis for standard setting.

Changeability of Standards

Standards change because of changes in (1) the technology with which we are able to monitor the environment, (2) our scientific understanding of micro- and macrocomponents and dynamics of the environment, and (3) public concerns for health, safety, and welfare as well as for the quality of the environment.

As shown in Table 4.1, standards that address particular environmental issues typically involve consideration of a range of parameters. For example, water quality standards applicable to public water supplies have typically focused on specific physical, chemical, and biological parameters of relevance to human safety and health. Included are standards for turbidity (i.e., the murkiness of water caused by suspended material that can decrease the efficiency of chemical disinfection), coliform bacteria (i.e., indicators of potential pathogens derived from fecal material), and a relatively small number of chemical contaminants that directly influence the potability of water. In recent years, increasing attention has been given to expanding the number of chemicals considered because of our rapidly expanding capacity to measure chemical contaminants and because of our growing concern about and our knowledge of the health hazards of chemicals. Also, increasing attention is being given to ecological phenomena such as bioaccumulation, which is the increase in concentration of toxic chemicals in the living tissue of fish and other aquatic wildlife, a development driven by public health considerations but also greatly influenced by our increasing knowledge about diverse and complex processes that characterize aquatic ecosystems.

Historically, noise standards tended to focus primarily on human comfort or safety. More recently, increased attention has been given to the potential effects of noise on wildlife.

Standards applicable to hazardous waste continue to give emphasis to the amount and type of waste and to specific operational measures and technologies for controlling their handling, transportation, and disposal. However, the issue of social equity, especially with respect to the inequitable distribution of risks associated with disposal technologies, is forcing consideration of a wide range of additional social factors that will have to be factored into the standard-setting process in addition to traditional physical, chemical, and biological factors.

Compatibility of Standards

The use of an environmental standard for impact assessment must be guided by the specific objective of that standard—an objective that is

TABLE 4.1 Examples of Parameters and Criteria Relevant to Key Environmental Issues

Environmental issues of concern	Relevant parameters and criteria
Air quality	Kinds and concentrations of chemicals
	Public health
	Vegetation and wildlife
	Corrosion of structural materials
	Acid deposition
	Social equity
Water quality	Physical, chemical, biological factors
	Human health
	Human recreation
	Agriculture and industry
	Wildlife habitat
	Ecological stability and diversity
	Bioaccumulation
	Social equity
Noise	Amplitude, frequency, and duration
	Temporal and spatial variation
	Physiological and psychological response
	Vibrational effects on unconsolidated soils
	Wildlife
	Social equity
Historic sites	Age
	Historic events/persons
	Architectural significance
	Educational opportunity
Pesticides	Lethality for target species
	Lethality for nontarget species
	Persistence in environment
	Degradation products
	Social equity
Radiation	Type and levels
	Effects on vegetation and wildlife
	Acute and chronic effects on humans
	Social equity
Hazardous waste	Amount and type
	Health and safety hazard
	Discharge of leachates or combustion products into surface and groundwaters
	Social equity
	Effects on vegetation and wildlife
	Effort of generator to reduce waste
Resource use planning	Critical areas/resources
	Multipurpose use of resources
	Sustainability

typically coupled to the legally permitted or promoted use of a particular resource. For example, dissolved oxygen and color standards may be applied to a particular reach of river to promote fisheries in that reach. Different standards for the same parameter may be applied elsewhere along that same river to promote other objectives, such as human recreation or potability. Do such human objectives totally encompass the environmental functions, components, attributes, and processes that constitute that river? The question of the compatibility of standards arises because conflicting objectives may be applied to the same resource.

A noise standard may be enforced in a particular area for the express purpose of insuring human comfort levels. However, a standard relevant to human comfort is not relevant to the comfort of a wildlife species that may be using that same area during its reproductive cycle.

Environmental standards, which typically represent relatively narrow human objectives, do not reflect a condition that is somehow "ideal" for the environment as a whole. To base impact assessment only on the question of whether or not a certain impact is within or exceeds a certain standard, therefore, constricts the meaning of the concept of environment to that of a single-purpose human resource. Although the examination of relevant standards is important in the evaluation of impacts, it is equally important that impact assessment employ a more comprehensive perspective.

Scientific Basis

Environmental criteria and, ultimately, standards are basically derived from experimental or *in situ* observations. Given the diversity of environmental parameters, the identification or selection of particular parameters for laboratory or field study presumes some degree of pre-existing concern or knowledge.

For example, researchers know that the interspersion of vegetative types is an important factor governing the population densities of various bird species. This knowledge, based on controlled field experiments as well as on professional observation, becomes the basis for establishing specific criteria for an ideal habitat for those species, including the presence of vegetative species, height of vegetation, distances between over- and understory, and surface water supplies. Implicit in the use of such criteria is that the relevant bird species is used as an "indicator species" of environmental quality, that is, as an integrative measure of the environmental impact of changing vegetative patterns throughout project development.

Whereas some criteria are clearly based on the use of "indicator species" or of species specifically selected because of our detailed knowledge and understanding of them (e.g., the U.S. Fish and Wildlife Service Habitat Evaluation System), other criteria give preeminence to certain species not because of our detailed knowledge of them, but because we recognize that they are more sensitive to environmental insults than others.

The LC_{50} of a chemical for a particular test organism is the ambient concentration of that chemical that is lethal for 50% of the exposed population. Those individuals in the population that are not killed at this concentration are more tolerant of the chemical and those that are killed are less tolerant, a fact that reflects the inherent diversity within any population. In an effort to provide some degree of safety for the less tolerant individuals in a population, the U.S. Environmental Protection Agency has suggested that, in some instances, a criterion must be established that is 1/100th of the LC_{50} whereas, in other instances, the criterion may be established as 1/10th or 1/20th the LC_{50}. Such differences in the safety margin of exposure (as expressed in the criterion) are based on the relative importance we give to the roles that individual species play in the ecosystem.

The examples just discussed demonstrate that, although criteria and standards typically involve data and information derived from scientific experimentation and observation, they also include judgments influenced as much by our lack of knowledge as by what we do know or think we know. An example of the degree of our ignorance about the public health effects of chemicals is included in Table 4.2. This table summarizes a recent evaluation of the adequacy of data pertaining to the health effects of chloroform, a chemical that has been in consistent use for 140 years. As indicated by this table, despite the long use of this chemical in medicine, data on the direct health effects of chloroform on human beings are essentially negligible. What we know about the possible health effects of chloroform on humans (as for most other chemicals) must be extrapolated from what we know about its effects on animals.

Another instance in which criteria may be severely flawed by our lack of knowledge is the application of epidemiological data on chemical exposure for the purpose of setting human health standards. As discussed in greater detail in Chapters 13 and 17, epidemiological studies of chemical effects on humans have been based largely on information derived from workforce populations. However, as shown in Figure 4.2, in the United States, females have begun to be represented equitably in the national work force only very recently. Thus, historical epidemiological data related to exposure to workplace chemicals and to chronic health effects are largely reflective of the male experience. Since health standards related to chemical exposure are based on these historical epidemiological data, many such standards may have primary relevance only to males.

TABLE 4.2 Adequacy of Database on Health Effects of Chloroform[a,b]

Health effects	Routes of entry		
	Oral	Inhalation	Skin absorption
Human data			
Lethality	+	−	−
Acute toxicity	−	+	−
Intermediate toxicity	−	−	−
Chronic toxicity	+	+	−
Developmental toxicity	−	−	−
Reproductive toxicity	−	−	−
Carcinogenicity	+	−	−
Animal data			
Lethality	*	+	−
Acute toxicity	*	+	+
Intermediate toxicity	*	+	−
Chronic toxicity	*	−	−
Developmental toxicity	+	*	−
Reproductive toxicity	+	+	−
Carcinogenicity	*	−	−

[a] After Oak Ridge National Laboratory, Toxicological Profile for Chloroforms, under DOE Interagency Agreement Number 1425-1425-A1.
[b] (−) Indicates no data available; (+) indicates some data available, but they are inadequate; (*) indicates sufficient data available to assess health effects.

RELEVANCE TO IMPACT ASSESSMENT

The use of standards in the identification and evaluation of project impacts has sometimes been referred to as "a shortcut to impact assessment." After all, standards typically provide a clean measure of consequence—if the degree of environmental impact is within a standard, developers presume that the impact is negligible.

As implied by the preceding discussions, this approach to impact assessment is dangerous because standards reflect interests, concerns, and knowledge that are essentially limited with respect to the complexity of the total environment. Existing standards must be used in the identification and evaluation of environmental impacts, but they do not circumscribe the assessment process. They are guides; they are not constraints.

With respect to the practical conduct of impact assessment, the following guidelines are suggested in an effort to maximize the usefulness of standards without constraining the environmental impact assessment process.

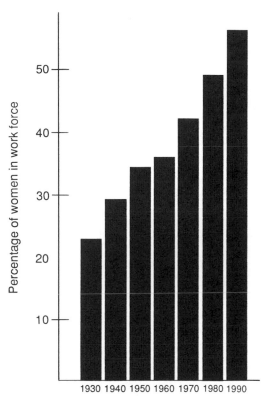

FIGURE 4.2 Civilian labor force rates for women 16 years of age and older. After U.S. Department of Labor, Bureau of Labor Statistics, *Handbook of Labor Statistics* 1993.

1. Identify all criteria and standards that apply to physical and social environmental components and dynamics that are possibly affected by the proposed project development.

2. Review the rationale that underlies each criterion and standard, giving particular attention to (a) the purpose of the criterion or standard with respect to resource use and quality and (b) limitations (e.g., with respect to resource use, natural variation of relevant parameters within the region, and climatic factors such as temperature that can influence the relevance of the standard or criterion).

3. Through liaison with relevant legal, technical, and scientific authority, (a) demonstrate the relevance of criteria and standards to each phase of the proposed project (i.e., early planning through construction, and operation and maintenance phases) and (b) identify those criteria and

standards (if any) that may be exceeded under specified conditions of frequency, magnitude, duration, and geographic extent.

4. Implement rationales and protocols for the documentation of baseline conditions (i.e., ambient, preproject conditions) relevant to quantitative criteria and standards, with special emphasis on (a) standard analytical methods, (b) location of sampling points, (c) sampling requirements (e.g., sample volume, frequency, timing), and (d) statistical analysis of data.

5. Establish precise protocols by which predictions of environmental impact relevant to established criteria and standards will be made and compared with previously documented ambient conditions (e.g., mass-balance calculations, mathematical or physical modeling, environmental monitoring during project development). Reviewing these protocols with relevant legal authorities prior to their implementation is important.

These five steps should be taken at the earliest possible time of the assessment process (e.g., during the "scoping process;" see Chapter 6) to allow the most cost-effective management of effort (as measured in time and cost) and to reduce the risk of an unexpected diversion of major effort away from the overall assessment effort to respond to unforeseen problems related to environmental criteria and standards.

CHAPTER 5

RESOURCE INTERACTIONS

Impacts on individual components of the environment that occur as a direct result of project-related activity are "primary," "direct" (Chapter 1), or "nonsystemic" impacts. For example, the dredging of bottom muds directly results in the removal of benthic organisms from a riparian habitat, just as clear-cutting directly results in the loss of vegetative biomass from a forest. However, most impacts on individual components of the environment occur as results of impacts on the dynamics of the environment and, in this sense, may be described as systemic impacts or "secondary," "second order," or "indirect" impacts.

A practical approach to defining and evaluating systemic impacts that may lead to numerous indirect impacts requires consideration of the resource interactions that may occur within terrestrial, aquatic, and wetland resources. These interactions may be broadly described as

- biotic and abiotic interactions,
- species interactions, and
- physical and social interactions.

BIOTIC AND ABIOTIC INTERACTIONS

Although we do not understand all biotic (living) and abiotic (nonliving) interactions that underlie what we perceive as the environment, certain of these interactions are understood well enough that they can be given particular attention in the process of impact assessment. Concepts and phenomena that denote some of these key interactions include

- habitat,
- niche,
- biogeochemical cycling, and
- ecological succession.

Habitat

Although habitat is often referred to as the address of an organism (i.e., where a specific type of organism lives; where one would expect to find it), considering habitat a specific place where physical, chemical, and biological factors and processes are suitable for a specific species is more meaningful for the purpose of impact assessment. Depending on the species considered, physical factors of its habitat might include such parameters as temperature, relative humidity, photoperiod, substrate type, and availability of water. Chemical factors might include the concentration of oxygen, pH, and availability of mineral nutrients. Biological factors might include the population density of predators, the degree of interspersion of vegetative types, and the availability of prey.

As a "place" categorized by a specific physics, chemistry, and biology, project-related activities may alter habitat by (1) removing that "place," either partially or entirely; (2) introducing it into an area where it was not previously present; or (3) altering its physical, chemical, or biological attributes.

For example, extensive clear-cutting may remove shrubs of various heights from an area that, because of the particular interspersion of those shrubs with open spaces and trees, provides ideal habitat for a particular bird population. The change in such biological factors in the project area may constitute a sufficient change in habitat that the specified bird population can no longer use the area. The direct impact is on vegetative diversity, but the indirect (or systemic impact) is on the avian species and on all the other features of the environment contributed to or affected by that species. Alternatively, a project may require the dredging of a stream, with the consequent direct removal of muds that provide appropriate substrate for certain species of benthic (bottom dwelling) organisms.

These examples, demonstrating the removal of a type of habitat with the consequent loss of a particular species, are also examples of the substitution of one type of habitat with another. Thus, in place of a habitat categorized by an interspersion of shrubs, open areas, and high trees, we substitute one best categorized as a grassland. Instead of a muddy benthic habitat, we substitute one categorized by a rock and gravel substrate. Depending on the species considered and the degree of change in habitat parameters, a change in habitat may generate a change in species.

Whether or not a particular area becomes unsuitable for a particular species depends on the type and degree of change in the habitat of that species, that is, in the physical, chemical, and biological factors that define that habitat. In addition, we must consider the tolerance of a species for change. For example, the grasshopper sparrow is severely restricted in its habitat to grassland, whereas the short-tailed shrew is able to utilize a variety of habitats, ranging from grassland to a high-tree habitat. Species of Salmonidae (salmon family) typically require high concentrations of dissolved oxygen (e.g., > 5 ppm), whereas species of Cyprinidae (carp family) are able to withstand much lower concentrations of oxygen.

Habitat requirements are not necessarily static. Depending on the species, habitat requirements may vary over time and in relation to the life stage of individual organisms. For example, habitats suitable for reproduction may be quite different from habitats that are used by an organism during nonreproductive periods. Habitats used during a seasonal migration may vary greatly from habitats used before or after that migration.

In many impact assessments, obtaining lists of plant and animal species that are known or likely to be present in a general geographic area is relatively straightforward. Such lists may be available through publications or may be compiled from the knowledge and experience of persons familiar with that area. However, an assessment team must take a more laborious approach, using various types of field studies to confirm the presence or absence of specific species (especially legally protected species) or to identify habitat types that suggest the presence or absence of species.

Niche

Whereas habitat may be described as an organism's "address" (i.e., where it lives), niche may be described as the organisms's "profession" (i.e., what it is doing in its habitat). Historically, professional ecologists have given somewhat different emphasis to various aspects of the niche concept. Some individuals, for example, utilize "niche" to refer specifically to what the organism does—they give emphasis to the *functional role* of the organism. Others utilize the concept to describe different *subdivisions of the physical environment* (i.e., habitat or, more precisely, microhabitat) in which the organism acts out its role(s). The former approach typically generates descriptions of ecological niches in terms of dynamic processes that are carried on by the organism; the latter generates descriptions of the physical and chemical factors of the environment that give a competitive edge to one species or another. The more contemporary view involves both approaches, that is, the functional role of an organism in the ecosystem as well as its position in time and space.

For purposes of impact assessment, giving priority to the functional aspects of the niche concept is usually most useful. For example, basic niches (in a very general, functional sense) in ecosystems include (1) primary production, (2) consumption, and (3) decomposition. Each of these "niches" may be defined by a certain type of work that must be done in ecosystems. The myriad organisms in various ecosystems, despite their apparently infinite diversity in shape, color, size, and behavior, are simply differently appearing and differently behaving machines by which ecological work is accomplished.

1. **Primary producers** are organisms that transform the radiant energy of sunlight and inorganic nutrients into living materials (i.e., they carry on photosynthesis). The only requisite "design specification" for such a transformer is that it have chlorophyll. Thus, all so-called "green plants" act as primary producers, including pine trees and microscopic, single-celled plants floating in the ocean. The ecological niche for both is the same. Each is doing the work of transforming the inorganic into the organic.

2. **Consumers** are organisms that cannot transform sunlight energy into the chemical energy of organic molecules. They, therefore, must consume preformed organic materials and transform these into the living substance of their own bodies. *Herbivores* transform the chemical energy of plant tissue into the chemical energy of their own animal tissue. *Carnivores* transform the chemical energy of other animal tissue into their own chemical energy. *Omnivores* transform the chemical energy of both plant and animal tissue into their own chemical energy. *Detritivores* transform the chemical energy locked up in dead, organic remains into their own chemical energy.

The niche of consumption, then, allows the ecologist to see an ecological equivalence (in function) in a cow and a grasshopper, because both do the work of herbivory; in a wolf and a bass, because both do the work of carnivory; in a carp and a human, because both do the work of omnivory; and in a snail and an earthworm, because both do the work of detritivory. Of course, the biological differences between members of these respective pairs are immense but, ecologically, each is the functional equivalent of the other.

3. **Decomposers** are organisms that accomplish the disintegration and mineralization of organic materials into inorganic materials. This process releases into the environment those inorganic nutrients previously locked up in the complex molecules of life. Decomposers include bacteria, fungi, earthworms, beetles, and any number of other organisms that transform organic into inorganic materials. Ecologically important is not who does the work but that the work is done.

These three basic ecological niches give emphasis to a particular perspective of organisms in general, namely, that whatever else organisms are, they are machines that do work. Moreover, they are particular kinds of machines. Organisms are transformers—machines that take in energy and matter of one kind and transform it into energy and matter of another kind. Because three basic kinds of work must be done in an ecosystem, essentially three kinds of living transformers exist. Each type of transformer can be found in aquatic, terrestrial, and wetland ecosystems. However, these organisms manifest different biologies, according to the constraints of habitat that are peculiar to each.

As depicted in Figure 5.1, one way to view these three basic types of ecosystem is to envision each as a specific combination of niche functions and of the various biotic and abiotic factors of habitat. Although the basic niches of primary productivity, consumption, and decomposition are common to the three basic types of ecosystem, habitat varies from one to another, thus providing the physical and temporal space in which different populations can perform their interdependent functions.

Biogeochemical Cycling

One of the more comprehensive examples of resource interactions is the model of biogeochemical cycling that includes the total dynamic relationships between the lithosphere (surface layers of earth), the atmosphere, the hydrosphere (surface and groundwater), and the biosphere (the total life on earth). Although such a global perspective underlies current serious concern over such issues as ozone depletion and global warming, the practical requirements of impact assessment cause us to focus on aspects

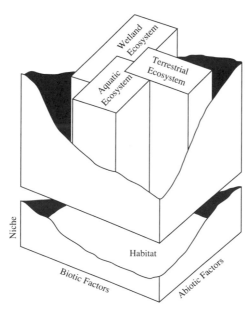

FIGURE 5.1 Distinct but interconnected ecosystems composed of specific types of habitat and ecological niches.

of biogeochemical cycling that are of immediate and relatively local (or regional) concern.

One aspect of biogeochemical cycling of particular importance in tropical and subtropical rainforests is nutrient recycling, the dynamic process by which essential plant nutrients (including carbon, nitrogen, phosphorus, and various trace nutrients such as heavy metals) are first incorporated by primary producers into their vegetative biomass and subsequently are released into the soil through decomposition of that same vegetative biomass. Because of the relatively high temperatures and moisture in tropical and subtropical rainforests, the rate of decomposition (mineralization) of dead organic material (detritus) is very high. Thus, nutrients trapped in dead vegetation are rapidly released to the soil and therefore can be rapidly taken up again by living vegetation. This rapid cycling means that, at any instant, most of the nutrient pool in a rainforest is locked up in living vegetation and relatively little is present in the soil, in contrast to the situation in a temperate forest in which, because of much lower rates of decomposition, most of the nutrient pool (in the form of incompletely mineralized detritus called humus) remains in the soil. Clear-cutting and removal of trees in a tropical or subtropical rainforest is, therefore, essentially the removal of plant nutrients—a consequence that has long-term systemic ramifications in terms of enhanced soil erosion and desertification.

Projects resulting in release of acidic discharges to the atmosphere (e.g., power plants) may also affect nutrient cycling because, on fallout to the ground, the hydronium ion (the positively charged hydrogen moiety that causes acidity) can differentially displace other positively charged trace nutrient ions (e.g., aluminum, cadmium, chromium, zinc) that are adsorbed to soil particles and, thereby, can cause the leaching of essential plant nutrients out of soils. Acidic deposition may also influence nutrient cycling in aquatic systems by increasing acidity to a level that is inhospitable to aquatic bacteria that mineralize organic matter. Without mineralization and its subsequent release of plant nutrients, aquatic primary producers are not able to survive unless they are fertilized by other means.

Ecological Succession

Ecosystems have their own life history. Deep, nutrient-poor, biologically inactive lakes may ultimately become relatively shallow, nutrient-rich, biologically active ponds. Similarly, grasslands may evolve through low and high bush lands to low and tall forests. The specific phases through which their life histories pass or at which they stop basically depend on the overall climatic conditions under which ecosystems exist, as well as on other factors, not the least of which is human activity.

The predictable, orderly changes in plant and animal communities are referred to as ecological succession. Essentially, ecological succession is driven by those physical and chemical changes that communities make in their environment as they unwittingly prepare that environment for subsequent, different communities (which will do likewise), until the ecosystem reaches a long-term stability with respect to its encompassing climate.

As a natural phenomenon, succession cannot be stopped except by the expenditure of energy. When climate dictates that grassland will turn into forest, the grass must be mowed (or grazed) and saplings cut (or debarked) to keep it grassland. For the purposes of impact assessment, we must distinguish between (1) project-mediated changes in habitat and (2) changes mediated by associated animal populations and succession.

For example, as plant succession proceeds from early pioneer stages through developmental stages, habitat changes will result in the loss of certain animal populations. A particular project may speed up the loss of such animals by removing habitat more quickly than would otherwise occur through succession. A project may also prevent the loss of such animals by keeping the project area in an early phase of succession.

SPECIES INTERACTIONS

Whether by removing habitat in the project area (e.g., through excavation or clear-cutting), by introducing new habitat into the project area (e.g., through purposeful vegetative landscaping or introducing new substrate), or by altering existing habitat (e.g., increasing the temperature of water by removing overhanging vegetation), projects can easily bring about changes in a biological community. Such changes may be manifest in changes in population densities (increases or decreases) and, in the extreme, changes in species (exclusions or introductions). Long- and short-term changes in species should be evaluated with respect to species interactions to determine potential systemic impacts of project development.

Species interactions, including interactions among terrestrial and aquatic species, and microscopic and macroscopic species, may be described in terms of the following general typology, which is based on the possible outcomes of an interaction between two different species:

1. **Mutualism:** Both species are benefitted by the interaction
2. **Predation:** One species is benefitted, while one is inhibited
3. **Competition:** Each species is inhibited by the other
4. **Commensalism:** One species is benefitted, while the other is not affected

5. **Amensalism:** One species is inhibited, while the other is not affected
6. **Neutralism:** Neither species is affected by the interaction

The practical applications of this typology to impact assessment (as opposed to ecological research) are best summarized by certain questions that should be addressed by the assessment team.

1. Will the removal of a particular kind of habitat throughout the project area result in the exclusion of a predator whose prey will therefore be able to proliferate? If so, what are the possible consequences, to the physical and the social environment, of such an unchecked proliferation?
2. Will the introduction of a new species, because of new habitat created as a result of project development, result in an increase in population densities of a commensal species? What might such an increase in population density mean in terms of the commercial use of this resource?
3. Will the removal of an amensal aquatic species, because of an increase in stream temperature caused by the removal of overhanging vegetation, result in a population increase in its amensal partner? What are the possible consequences to the fisheries of this stream?

Although these questions do not exhaust all the possibilities, they are sufficient to underscore the importance of viewing different kinds of organisms as interlinked not only with the physical and chemical attributes of their environment, but also with each other. Project-mediated direct or indirect effects on one population may have significant systemic effects on another.

PHYSICAL AND SOCIAL INTERACTIONS

Although we do not yet have a full understanding of all the varied aspects of the interplay of personal, interpersonal, and institutional life and the physical environment, viewing that interplay as a highly dynamic phenomenon in which society not only shapes its environment but is also shaped by its environment is instructive. In this sense, then, environmental processes have their social cognates, and vice versa. As practiced for almost a quarter century, impact assessment has tended to focus on social impacts on the environment; it has ignored the question of environmental impacts on people.

A balanced assessment of project-mediated impacts on the dynamics of physical and social interactions is perhaps most practicably attained at this time by focusing on certain phenomena that may be described best as "environmental risks and benefits." For example, much attention has been given in the United States to the preservation of wetlands. Beginning with growing concern over the loss of wetland habitat, the commitment to wetland protection has been increasingly strengthened by our growing understanding of the many environmental functions that wetlands perform in addition to providing wildlife habitat. However, many wetlands also serve as collection areas for runoff containing heavy metals. Many of these areas also serve as habitat for disease vectors as well as "wildlife." Many also present safety risks to children. To say that a wetland might contaminate drinking water, or present health or safety risks to humans, is not to say that a wetland does not perform other desirable functions. However, we must take a comprehensive view of the possible benefits *and* the possible risks of a particular wetland.

Basic issues to which a balanced assessment of environmental benefits and risks must be addressed include (but are not necessarily limited by):

1. **Human health, safety, and welfare:** Includes consideration of physiological, physical, and psychological harm; includes impacts with respect to disease, chemicals, and physical agents such as noise and radiation
2. **Subsistence:** Focuses on the physical and cultural requirements of indigenous peoples
3. **Economic well-being:** Focuses on distributive as well as nondistributive aspects of monetary earnings
4. **Recreational opportunities:** Includes active and passive recreation as defined by local population as well as by societal norms
5. **Educational opportunities:** Includes institutional and noninstitutional (or *ad hoc*) forms
6. **Self- and social image:** Focuses on impacts on values and self-esteem of affected populations.

CHAPTER 6

THE ASSESSMENT PROCESS

Although many differences exist in the legislation and executive orders that govern the conduct of impact assessment in different nations, a feature common to all is that **the goal of impact assessment is, first and foremost, to improve the decision-making process by which projects are planned, designed, and implemented.**

The scientific and technical difficulties inherent in impact assessment often obscure the fact that the objective of impact assessment is not the furthering of scientific knowledge, but the improvement of decision making by forcing consideration of a range of possible consequences of human actions that, historically, has been ignored in favor of political expediency and narrow economic interest. Therefore, examining the practical consequences of environmental impact assessment as an integral part of decision-making processes before focusing more intently on its scientific and technical machinations is essential.

OVERVIEW OF DECISION MAKING

Any formal decision-making process (e.g., governmental decision making with respect to undertaking a proposed project) may be viewed as consisting of several basic components. These components of decision making include its (1) experiential base, (2) prediction system, (3) value system, and (4) selection system (see Figure 6.1).

The *experiential base* of a decision-making process is the sum of all information, data, and knowledge pertinent to that process. For a governmental agency, this base consists of its legal authorization and all previous actions and experiences as well as protocols previously established in carrying out the mandates of its authorization. This experiential base defines the possible actions that may be undertaken, as well as the goals and objec-

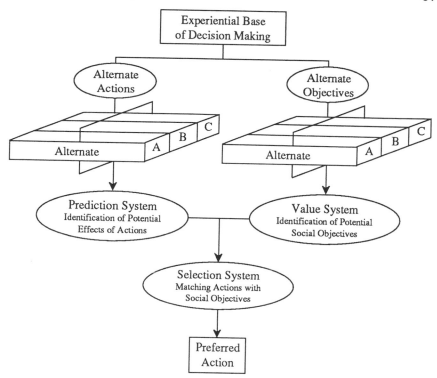

FIGURE 6.1 The decision-making process as a series of interrelated prediction, value, and selection systems.

tives that are to be considered appropriate. Which of various possible actions and which of various objectives will, in fact, be pursued at a given moment is the question that initiates decision making.

From the perspective of decision making, whether conducted by an individual, a government, or corporations, good decisions are those that result in the realization of a selected future; bad decisions are those that do not result in the realization of a selected future. At least some portion of the difference between a good and a bad decision is, therefore, the difference between a good and a bad *prediction system*. A valuable prediction system in decision making is a tool that allows us to relate our present actions accurately to the future consequences of those actions.

Contemporary technobureaucratic societies may be described as essentially pluralistic, that is, they are characterized (despite national pronouncements to the contrary) by diverse values and attitudes. In such societies, the *value system* promulgated by one institution (e.g., a religious institution) may often conflict with that promulgated by another (e.g., a corporation). Because values are important factors in human motivation to undertake specific actions, the identification of appropriate values is a complex process in pluralistic societies. Some decision makers, of course, must be concerned with the values of only a relatively small number of people (e.g., a corporate board of directors); others have larger constituencies (e.g., a political party). Decision makers involved in environmental decision making, as defined by legislation or executive orders that mandate environmental impact assessment, must interact with very diverse groups, organizations, and agencies and, therefore, with a large number of diverse values.

The manner in which pluralistic values and predicted consequences of actions are integrated, and in which preferred actions and, therefore, alternative futures are selected, constitutes the *selection system* of decision making. This phase of decision making may involve complex institutional and public interaction, such as a referendum or a public hearing. This phase may also be left to the judgment of an individual or a small select group, as in the selection of technical specifications for construction materials and supplies.

IMPACT ASSESSMENT AND DECISION MAKING IN PROJECT DEVELOPMENT PROCESS

A good rule of thumb is that any disagreement about what a law says occurs because of what the law does not specifically say. The scope of the United States National Environmental Policy Act (NEPA) is certainly more easily approached in this manner. For example,

- NEPA does *not* say that the historical experiential base of any federal agency shall be given absolute and exclusionary priority in planning current projects and programs.
- NEPA does *not* say that any particular protocol or system for making predictions shall be given absolute and exclusionary priority.
- NEPA does *not* say that any single social value, or any single criterion for selecting among conflicting social values, shall be given absolute and exclusionary priority in decision making.

What NEPA does say with respect to these components of decision making is that federal agencies, in fulfilling their legal obligations,

- will utilize interdisciplinary approaches,

- will insure an integration of natural and social sciences and environmental design arts, and
- will insure that unquantifiable environmental amenities and values are given appropriate consideration in addition to economic and technical considerations.

Considering what NEPA does *not* say as well as what it does say, an American citizen can expect the following features to be part of current decision making and planning by federal agencies whose actions may significantly affect the environment.

1. Regardless of an agency's previous experiential base for decision making, that base should be supplemented with interdisciplinary concerns.
2. Regardless of the prediction system and techniques formerly employed for planning, design, and implementation of projects, those systems and techniques should be supplemented with methodologies and procedures that allow for the consideration of unquantifiable environmental amenities.
3. Regardless of originally authorized objectives and regardless of value selection criteria historically employed by a federal agency, those values and criteria should be supplemented with the broad environmental concerns and goals of the nation as set forth in NEPA.

This revolution in agency decision making has resulted in very real frustration and problems, but was nevertheless undertaken because "previous practice was inadequate—agencies' planning and decision-making procedures had ignored or misjudged significant environmental impacts and would continue to do so" (U.S. Council on Environmental Quality, 1975, *Sixth Annual Report of the Council on Environmental Quality,* U.S. Government Printing Office).

Regardless of the many problems and inefficiencies that have occurred as a result of incorporating broad environmental concerns into the governmental bureaucracy of decision making, environmental impact assessment is increasingly becoming a basic tool for decision making throughout the world. As should any tool, environmental impact assessment should be evaluated by the results actually achieved by its use.

Of course, only one result is expected from the use of environmental impact assessment—decisions! Good impact assessments help make good decisions. If impact assessments are conducted so their contribution to decision making in project development is irrelevant, through their failure to make specific and timely recommendations or to consider the environment from a comprehensive perspective, then impact assessment, regardless of its scientific or technical merit, is also irrelevant.

Decision Making in Impact Assessment Process

The various phases of the impact assessment process have their own decision-making requirements. Although types and examples of decisions (and guidelines for making them) regarding the identification and mitigation of impacts are discussed in the subsequent sections of this volume, two phases in which the decisions to be made are of such importance to the overall assessment process must be given early and particular emphasis. These phases are best known as the "scoping process" and the "impact evaluation process."

Scoping Process

The scoping process begins at the very start of the assessment process and, depending on project-specific circumstances, may be extended. In essence, the scoping process consists of the early identification of environmental issues and concerns **likely to be significant,** toward which the assessment process should be directed. The goal of the scoping process is minimizing inefficiencies of the assessment process by focusing attention on realistically probable rather than improbable issues and concerns. Toward this end, the scoping process typically assumes an overall managerial function—defining and setting assessment objectives, making personnel assignments, and determining schedules and budgets. Ideally, the advantage of the scoping process is that, by minimizing inefficiencies, it can direct maximum effort to the analysis of matters of real consequence. However, a possible disadvantage is that the process can set narrow constraints on assessment efforts and thereby increase the likelihood of oversight of unforeseen issues.

The key to a successful scoping process is, of course, the personnel who conduct it. Given the practical constraints of budgets, schedules, and the availability of personnel within which every assessment team must work, prescribing precisely the professional occupations and disciplinary expertise that must be represented by the personnel responsible for "scoping" an impact assessment is pointless. However, describing some basic goals is useful.

1. Because the assessment process is essentially an interdisciplinary effort, and because it must address the social and the physical environment, as well as their interactions, scoping should be conducted by personnel who represent a balance of physical science and social disciplines.

A scoping team composed essentially of wildlife biologists and engineers is hardly likely to recognize potentially important sociological ramifications of project development. Similarly, a team composed essentially of urban sociologists and anthropologists is hardly likely to recognize potentially important ecological ramifications of project development. Specialists in air quality are typically ignorant of kinship obligations that may define social interactions. Specialists in water quality are typically ignorant of the relationships between ecological succession and land-use patterns. Note that one of the major failings of impact assessment over the past 20 years has been the conspicuous absence of human health specialists among all the various physical and social specialists influencing the conduct of impact assessment, demonstrating that any historical imbalance in impact assessment has been (at least in the United States) in favor of the physical environment.

2. Academic disciplinary knowledge is important in the assessment process, but human experience is not confined by the compartmentalizations of knowledge that we call disciplines. Therefore, academic disciplinary understandings of environmental phenomena must be balanced with experiential understanding. For example, in some assessments in which impacts on an impoundment may be of primary concern, local fishermen who have fished that resource (whether for commercial, recreational, or sustenance purposes) may be far more knowledgeable than an academically trained limnologist or fish and wildlife biologist of events that can affect species of concern.

For the purpose of identifying possible impacts, as well as for the purpose of evaluating the significance of impacts, local experience—experience not sanctioned by academic credentials—can play a critical role, particularly in the assessment of direct and indirect impacts on social components and dynamics. Whereas all legally required impact assessments mandate (to greater or lesser degree) public involvement, too few assessment teams make direct use of the public (as opposed to paid consultants and contractors), especially during the scoping process. This approach increases the probability that site-specific information and local values and concerns will be overlooked, with consequent diminution of the relevance of the assessment.

3. Regardless of the legal authority that mandates an assessment process, the results of an assessment will be subject to some type of governmental review that, depending on the national authority, may include local, regional, or national authorities and agencies. By the time assessment findings become available for this review, significant efforts (in time and money) have been expended. At this point, serious deficiencies discovered during the review process become extremely problematic.

When possible, inviting the participation of governmental authorities in the scoping process is the wisest course so their jurisdictional interests and obligations can be identified precisely with respect to the proposed project- and site-specific conditions, attributes, and issues. Of course, for some governmental authorities to decline the invitation is not uncommon.

However, even under such circumstances, maintaining contact with these authorities throughout the scoping process and informing them of any unforeseen or special findings that fall within their jurisdictional purview is advisable.

Impact Evaluation Process

Having identified possible impacts of project development, evaluating those impacts with respect to their significance becomes necessary. Is this impact significant or not? Is it a significant adverse (or negative) impact? Is it a significant beneficial (or positive) impact? Performing this evaluation is necessary because decision makers must be informed about issues to which they should pay careful attention. No other task in the assessment process is more difficult.

Although the word "impact" has acquired a quasi-legal meaning of negativity among individuals having limited experience with the assessment process, impacts are merely "consequences" of proposed actions. These consequences are in and of themselves neither adverse nor beneficial, neither significant nor insignificant. Human judgment about these consequences, not the consequences themselves, determine their value.

For example, is the proposed conversion of borrow pits in the project area to wetland resources a beneficial or an adverse impact of project development? Whether beneficial or adverse, how does one determine its significance? From the perspective of waterfowl that might utilize the new wetland habitat, that wetland may reasonably be considered a beneficial impact of project development. From the perspective of various environmental functions typically associated with wetlands (e.g., recharge of groundwater aquifer, soil retention, tertiary treatment of land runoff), that wetland may also be considered beneficial. However, consider the perspective of parents in the local area. Might we not reasonably consider that wetlands also present safety hazards to small children? Might we not consider that wetlands serve as habitat for locally significant disease vectors (e.g., mosquitos, rodents)?

The fact is that **any one consequence (impact) of project development may present simultaneously any number of beneficial and adverse attributes.** As shown in Figure 6.2, valuation of impacts with respect to negativity or positivity must be done by the assessment team by considering project- and site-specific conditions of the physical and the social environment. Having made that determination, each adverse and beneficial impact must be evaluated for significance—an even more difficult task simply because no universally accepted measure of significance is available.

Lacking a predetermined measure of significance, our experience with impact assessment over the past 20 years suggests several criteria that the assessment team should consider in making their determination of signifi-

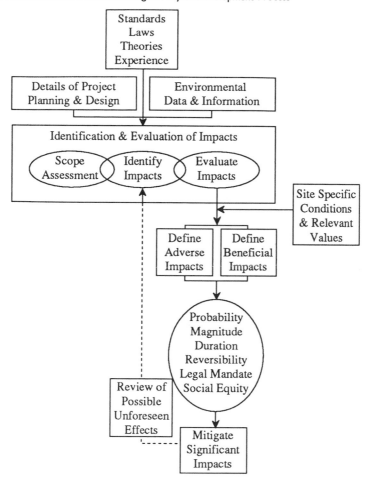

FIGURE 6.2 Key inputs and outputs related to the evaluation of impacts. Note that the broken arrow is a feedback mechanism requiring the review of proposed mitigation measures with respect to possible unforeseen consequences.

cance. None of these criteria should be considered inherently more important than any other. Their primary use is in helping assessors frame a rationale for making their determination and for communicating why they made the determination they did. These criteria include, but are not restricted to,

1. **Probability of occurrence:** Quantitative or qualitative estimate of the likelihood that the impact will occur; most estimates are given qualitatively using phrases such as "highly likely" or "possible but not very likely"

2. **Magnitude:** Quantitative or qualitative estimate of size or extent of the impact, for example, "a 30% reduction in the available habitat"
3. **Duration:** The period of time the impact, if it occurs, can be expected to last (e.g., decades or more, years, months, days)
4. **Reversibility:** Whether the impact can be reversed through human action or naturally
5. **Relevance to legal mandate:** The existence of local or national laws or international treaties that specifically promote or disallow this type of impact
6. **Social distribution of risks and benefits:** Whether this impact (whether adverse or beneficial) contributes to (or mitigates against) the equitable sharing of environmental risks and benefits

Although an impact may have a low probability of occurrence, this impact is not necessarily insignificant. For example, the spillage of 5000 gallons or more of a toxic chemical on an American highway is an event of extremely low probability. However, the potential magnitude and duration of the impact of a tanker accident in which its spillage resulted in runoff of 5000 gallons of a water-soluble toxic chemical into a surface drinking water supply serving 1 million people would be very great. Certainly public health officials and water supply managers would consider such an impact significant.

Similarly, a high magnitude impact is not necessarily a significant impact. For example, in a particular stream that supports a warm water fish species, dredging operations (e.g., for a new bridge) may result in a 100% loss of those fisheries. However, depending on the species involved as well as on other site-specific factors, the fisheries may be able to reestablish themselves within one season without any measurable environmental ramifications. In such an instance, the impact might be considered insignificant.

As these examples demonstrate, these criteria cannot be inserted into a formula that will automatically produce a finding of significance or insignificance. Criteria simply aid the assessor in making his or her judgment of significance on a case-by-case basis. This case-by-case approach—an approach that heavily depends on site-specific information and context—gives good assurance that judgments of significance will be relevant to the real rather than an abstract environment.

PART II

THE PHYSICAL ENVIRONMENT

CHAPTER 7

AQUATIC RESOURCES

Aquatic resources include

1. **Lentic resources:** Standing or "quiescent" freshwater systems such as lakes, ponds, and impoundments, sometimes referred to generically as lacustrine systems
2. **Lotic resources:** Running freshwater systems such as rivers, streams, and creeks, sometimes referred to generically as riverine systems
3. **Estuarine resources:** Areas where rivers empty freshwater freely into the sea and that are subject to tidal influence
4. **Groundwater resources:** Including water table and artesian aquifers

For purposes of impact assessment, considering that, with the exception of water table and artesian aquifers, these systems are characterized by a diversity of organisms that carry on primary production, consumption, and decomposition is useful. Although these niches (Chapter 5) are common to lacustrine, riverine, estuarine, and open-ocean ecosystems, the organisms that perform these functions vary among and within these systems depending on the physical, chemical, and biological constraints of their environment.

LACUSTRINE SYSTEMS

Key Dynamics

An example of how these three basic niches (or ecological functions) may be interrelated in lacustrine systems is shown in Figure 7.1. In this figure, the rectangles represent the basic types of energy transformation (Chapter 5) accomplished within each niche as well as the types of organisms

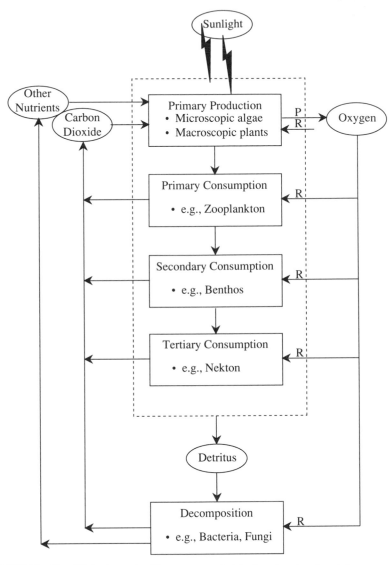

FIGURE 7.1 Simplified schematic of interconnected niche functions in a lacustrine ecosystem. Arrows represent the flow of matter, which is recycled by systemic dynamics. P represents the photosynthetic production of oxygen; R represents the respiratory consumption of oxygen.

that may perform those transformations. In this scheme, the radiant energy of sunlight is transformed by primary producers into the chemical energy of plant biomass. The chemical energy of plant biomass is then transformed by primary consumers into the chemical energy of their biomass which, in turn, is sequentially transformed into the biomass of a series of subsequent consumers. Of course, not all plant or animal biomass is consumed; much of it accumulates as detritus (dead organic material), which is the basic energy supply for decomposers. Decomposition results in the recycling of essential mineral nutrients (e.g., nitrates, phosphates) that, in conjunction with the carbon dioxide produced through the respiration of consumers and decomposers, are essential inputs to primary productivity.

The specific link between an energy source and the biotic species that directly or indirectly depend on that energy may be described best in terms of the various grazing and detritus food webs (Figure 7.2) that characterize a specific lacustrine (or other) ecosystem. Because these food webs depict the sequential transformations of energy achieved through consumption, they not only define the interdependence of biological species but also identify potential pathways by which toxic chemicals may accumulate in those species. For example, filter-feeding organisms (e.g., clams, oysters) derive their nutrition by filtering microbes (such as microscopic algae) and detritus out of water. If filtered materials are contaminated with even low concentrations of toxic chemicals, these toxic materials accumulate in the living tissue of the filter feeder and, as the animal continues to filter water, increase in concentration in that tissue. This process is an example of *bioaccumulation.*

Another example of bioaccumulation is the increase in the concentration of toxic materials in higher order consumers. This type of bioaccumulation, called *trophic* (or *bio-*) *magnification,* occurs because any organism

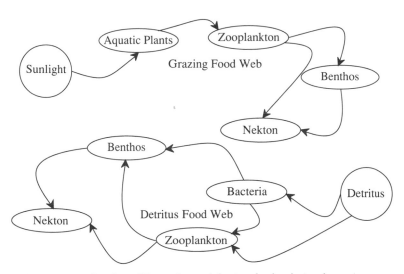

FIGURE 7.2 Examples of possible grazing and detritus food webs in a lacustrine ecosystem. Sunlight drives the dynamics of a grazing food web; the chemical energy of detritus drives the dynamics of a detritus food web. In most systems, these webs are linked by one or more species that participate in both webs.

transforms into biomass only a small amount of the energy actually consumed, the largest portion being spent in the chemical processes and behaviors associated with obtaining food, processing that food, and carrying on other necessary bodily functions. To increase biomass, therefore, a predator must consume relatively large amounts of prey; if individual prey animals are contaminated with relatively low concentrations of a toxic chemical, the predator, by consuming numerous contaminated prey, will become contaminated with relatively high concentrations of that same toxic chemical. Of course, this assumes that the toxic chemical is not eliminated as quickly as it is consumed.

Photosynthesis, the use of sunlight energy to combine carbon dioxide and water, is the chemical basis of primary productivity, producing oxygen as a waste product. In quiescent waters, this photosynthetically produced oxygen is typically the largest supply of dissolved oxygen on which all oxygen-requiring aquatic organisms depend. The available supply of dissolved oxygen (ignoring the amount contributed by the mechanical mixing of water with the atmosphere in wave action) is the difference between the gross amount of oxygen produced by the photosynthetic activity of chlorophyll-bearing plants and the amount consumed through their respiration. This difference is called the *net production of oxygen.*

Because the amount of dissolved oxygen in quiescent water depends on photosynthesis, and because the availability of the sunlight that drives photosynthesis decreases with depth, the amount of dissolved oxygen in lacustrine systems typically decreases with depth. This condition is manifested in the *compensation depth,* the depth at which the rate of the respiratory consumption of oxygen by chlorophyll-bearing plants is equal to their photosynthetic production of oxygen. Below the compensation depth, too little light is available for photosynthesis to provide a net production of oxygen. As other aquatic organisms continue to draw down an oxygen supply that is not replenished, these depths become oxygen deficient (*anaerobic*).

Of course, the compensation depth can vary significantly, even over short periods of time. For example, this level will be much higher in the water column on a cloudy day than on a sunny day. It will be lower in the water column during periods of low color or turbidity than during periods of high color or turbidity (which decrease sunlight penetration).

Daily and seasonal variations in compensation depth and the thermal profile of lacustrine systems in temperate climates place severe constraints on aquatic habitat. Figure 7.3 depicts a temperature profile of a temperate lake during a late-summer period. During this period, a layer of higher temperature water (*epilimnion*), which is relatively less dense, overlies a layer of lower temperature water (*hypolimnion*), which is relatively more dense. The two layers are separated by a third layer (*metalimnion,* also

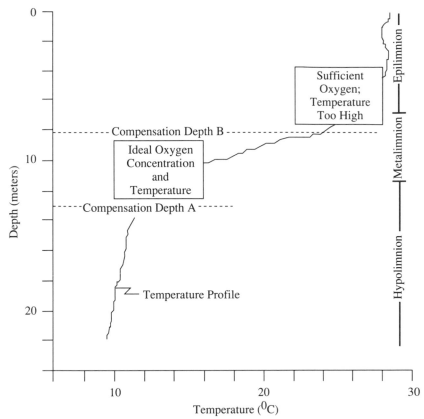

FIGURE 7.3 Changes in compensation depths with respect to the temperature profile of a lacustrine system can influence the survival of aquatic species with specific tolerances regarding temperature and the concentration of dissolved oxygen. As compensation depth moves from A to B, fish species may be forced to migrate upward to obtain sufficient oxygen, but suffer heat stress in the process.

called the *thermocline*) that is characterized by a significant decline in temperature with each meter in depth (usually $\geq 1°C/m$).

Figure 7.3 also depicts examples of compensation depths, demonstrating, for example, that a cold water fish species requiring relatively high concentrations of dissolved oxygen (>5 ppm) as well as a relatively cold temperature (<18°C) may not be able to survive. To obtain sufficient oxygen, it must move into water that is too hot and will suffer thermal death; to maintain appropriate temperature, it must move into water that is oxygen deficient and will suffocate.

The thermal stratification of temperate lakes not only defines available habitat but also results in a *thermal impedance to mixing* (Figure 7.4).

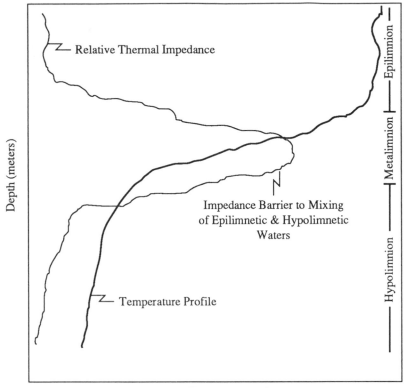

FIGURE 7.4 Relative thermal impedance, which can be calculated from the temperature profile of a lacustrine system, may result in an impedance barrier to the mixing of thermally stratified waters. Relative thermal impedance is the ratio of the density difference between water at the top and bottom of each successive 0.5 m of water to the density difference between water at 5 and 4°C.

If thermal impedance is strong enough to withstand the shear effect of surface wind, surface (epilimnetic) layers and bottom (hypolimnetic) layers will not mix. During this period, nutrients released through the decomposition of organic muds into the hypolimnion stay within the hypolimnion, that is, they cannot move into the epilimnion where the availability of sunlight would result in their utilization by epilimnetic primary producers. Therefore, during strong thermal stratification, the population density of surface layer primary producers such as phytoplankton will reach some maximum and then, because additional nutrients are not available, begin to decline.

When thermal stratification seasonally declines and surface and bottom waters are able to mix, surface waters become fertilized by the nutrients previously entrained in the hypolimnion and the population density of primary producers in surface waters increases. The seasonal mixing of surface and bottom waters is referred to as the overturn of the lake. Depending on depth and location, some temperate lakes overturn twice (e.g., spring and autumn); some overturn once or several times.

In addition to receiving nutrients through the decomposition of their own detritus, lacustrine systems also receive nutrients through land runoff, inflowing streams and groundwater, and, of course, human activity. Lakes that have very sparse concentrations of nutrients (and therefore support relatively little biological productivity) are referred to as *oligotrophic systems*. Lacustrine systems characterized by high concentrations of nutrients (therefore supporting the production of large amounts of aquatic biomass) are referred to as *eutrophic systems*. When the large amounts of nutrients contributed to a eutrophic lake are the result of human activity, we use the term *cultural eutrophication*. In some instances, large concentrations of nutrients do not result in enhanced biological activity, perhaps because of the presence of toxic chemicals or very acidic conditions. In such instances, the system is referred to as a *dystrophic system*.

The natural history of any lacustrine system is to proceed from a condition of lesser to one of greater concentration of nutrients (i.e., from oligotrophy to eutrophy), if only because nutrients are continually leached from surrounding land into the water column. Increasing eutrophication results in greater and greater amounts of biomass produced and deposited within the system, until the system finally becomes terrestrial. Cultural eutrophication greatly accelerates the process, compressing events that would, in the absence of human contributions of nutrients, possibly take thousands of years into a few decades or less.

As nutrient concentrations increase, the biomass of primary producers increases, as does competition among primary producers for sunlight. Typically, this competition results in dense mats of algae floating at the surface of the water. In turn, these dense mats effectively prevent sunlight penetration into the water column, causing the compensation depth to rise toward the surface. The shallow compensation depth and increasing oxygen demands of the decomposition of an enhanced detrital loading of bottom waters can quickly cause the water column to become essentially anaerobic, resulting in foul odors and a shift in species toward those with a high tolerance for low concentrations of oxygen.

The reason that foul odors are produced in anaerobic systems is that anaerobic decomposition is less efficient than aerobic decomposition. Aerobic bacteria (i.e., species requiring oxygen) are capable of transforming (through various steps mediated by different species) complex organic molecules such as proteins into minerals such as phosphates and nitrates, carbon dioxide, and water. This decomposition is often referred to as *mineralization* because it produces simple minerals, none of which is odoriferous. However, as conditions become anaerobic, the aerobic bacteria cease to function and anaerobic bacteria, to which oxygen is a poison, are activated. However, anaerobic bacteria cannot achieve complete mineralization. They achieve an essentially incomplete decomposition, producing as final products relatively complex molecules (e.g., aldehydes, ketones) that are typically highly odoriferous.

Most decomposition in lacustrine systems takes place within the first few millimeters of the top layer of bottom organic muds where, if hypolimnetic waters are aerobic, oxygen is still sufficient for aerobic decomposition. If this top layer of organic mud is covered with inert sediment (which will decrease the concentration of oxygen at the surface of the mud) or if hypolimnetic waters become anaerobic, the consequent partial decomposition characteristically performed by anaerobic decomposers is likely to result in an enhanced rate of thickening of the mud bottom. Conversely, the water column becomes more shallow. This "filling-in" process is accelerated by the increased detrital loading that accompanies eutrophication.

The *Index of Shoreline Development* (Figure 7.5) is a mathematical index that measures the relationship between the surface area of an impoundment and the length of its shoreline. If the index equals 1, the impoundment is circular—it has the shortest shoreline possible for a given surface area. As the index increases beyond 1, the impoundment has an increasingly sinuous shoreline. In general, the more shoreline per surface area (i.e., the more sinuous the shoreline), the greater the amount of soil nutrients that can be leached into the impoundment. Areas of shoreline that are more sinuous are therefore most likely to produce larger populations of primary producers and "fill in" with partially decomposed detritus more quickly than areas that are less sinuous.

Examples of Direct and Indirect Impacts

All examples of impacts in this and subsequent chapters are numbered to reference the chapter in which they occur, so various interconnected impacts may be cross-referenced appropriately.

Project activities resulting in impacts on a lacustrine system do not have to be conducted in the immediate vicinity of that lacustrine system, as evidenced by the following examples.

Example 7.1   Removal of vegetation, especially tall trees, along a windward azimuth toward a thermally stratified lake (Figure 7.6) will directly result in the removal of a physical obstacle to prevailing winds and,

Index of Shoreline Development

The ratio of the length of a lake's shoreline (including islands) to the circumference of a circle having the same area of that lake

Derivation: Let S = the length of the lake's shoreline, X = the surface area of the lake, and C = the circumference of a circle having an area equal to X. The Index of Shoreline Development (ISD) is therefore defined as:

$$(1)\ ISD = \frac{S}{C}$$

where (2) $\quad C = 2\pi R$ (definition of circumference of circle with radius R)

where (3) $\quad X = \pi R^2$ (definition of area X of circle with radius R)

Solve for R in Eq. 3 $\quad R = \sqrt{\dfrac{X}{\pi}}$

Substitute solution for R in Eq. 2 and solve for C

$$C = 2\pi \sqrt{\frac{X}{\pi}}$$

$$C = 2\pi \sqrt{\frac{X}{\pi} \cdot \frac{\pi}{\pi}}$$

$$C = 2\pi \sqrt{\frac{X\pi}{\pi^2}}$$

$$C = \frac{2\pi}{\pi} \sqrt{X\pi}$$

$$C = 2\sqrt{X\pi}$$

Substitute solution for C in Eq. 1

$$ISD = \frac{S}{2\sqrt{X\pi}}$$

Circular Lake: ISD = 1
Such a lake has the minimum length of shoreline per surface area

Subcircular or Elliptical Lake: ISD = 2+

Highly Configured Lake: ISD = 5 - 7

Extremely Configured Lake: ISD = 10 - 15

FIGURE 7.5 Derivation and relevance of the index of shoreline development as a measure of an impoundment's shoreline configuration.

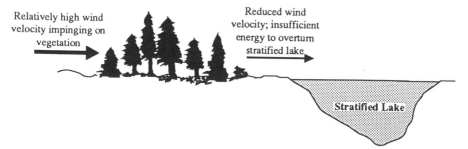

Relatively high wind velocity impinging on vegetation

Reduced wind velocity; insufficient energy to overturn stratified lake

Stratified Lake

FIGURE 7.6 The removal of tall trees, which can reduce the shear velocity of wind across the surface of a stratified lake, may lead to a wind-mediated overturn of the lake.

thereby, in a possible increase in wind velocity across the surface of that
lake. Depending on the increase in velocity and the degree of thermal impedance within the stratified water column, the lake could overturn with the
result that nutrient-rich hypolimnetic waters could fertilize nutrient-poor
epilimnetic waters. Depending on the amount of nutrient increase in surface
waters, population densities of primary producers (e.g., algae) could increase, causing color interference with recreational use (e.g., swimming),
odor and taste problems associated with drinking water supplies, as well
as changes in compensation depth due to the turbidity and color associated
with algal growth.

Example 7.2 Fugitive dusts and/or emissions from construction or
operations windward to a lacustrine system could fall out directly on its
surface or in its watershed. Depending on the toxicity of these materials,
they may have deleterious effects on key components of grazing and detritus
food webs that characterize the lacustrine ecology. Depending on their toxicity and chemical nature, as well as on the system's food webs, some of these
materials might be biomagnified, with consequent effects in higher order
consumers (including humans).

Finally, acidic emissions that precipitate in the watershed or on the
water resource may directly result in lowering the pH of the water, with
specific effects on pH-sensitive species. Of particular note is the possible
interference with the decomposition process (which accomplishes nutrient
recycling) because of the intolerance of many decomposing bacteria to low
pH (i.e., acidic conditions). Acidic deposition to watershed soils also can
result in leaching from those soils of various cations (e.g., aluminum, zinc)
which, on entry into a receiving lacustrine system, may exert toxic effects
on lacustrine biota or may cause changes in water quality that could affect
the preferred use of that resource.

Example 7.3 Blasting within the system's watershed or major earth-
moving may result, in the first case, in fractured artesian aquifers or, in the
second, in redirected flows of water table aquifers (see "Groundwater")
that serve as major groundwater sources to the impoundment. Subsequent
changes in water quality or quantity could have long-term, higher order
impacts on the system's ecology and attributes.

Of course, many activities or project design features conducted or
manifest in the vicinity of a lacustrine system may result in direct and indirect
impacts.

Example 7.4 Many types of development project (e.g., highways
and roads, housing) typically improve access to or increase the human use
of lacustrine systems, in the form of primary and secondary recreation. The
use of power boats, for example, can result in the stirring up of bottom

muds. Consequently, organic molecules previously sequestered in deep bottom muds can be released into overlying aerobic waters. This overturn will result in enhanced decomposition of those organic muds, and subsequent enhancement of eutrophication. The same effect may be realized by allowing the use of nonmotorized boats or swimming in shallow mud-covered bottoms that can be disturbed easily by the action of oars or poles and the movement of humans.

All these activities may also result in the release of previously sequestered toxic chemicals (e.g., heavy metals) to the water column, leading to a possible bioaccumulation of those toxics in food web species.

Example 7.5 Improved access to or use of lacustrine systems (Ex. 7.4) can also result in the disposal of sewage (human waste, husbandry wastes) within the watershed, the direct disposal of urine into the water by swimmers, the leaching of fertilizers applied to lawns and gardens within the watershed, and the disposal of lawn clippings and other vegetative debris in or near the water.

Impacts related to construction and maintenance activities conducted in the watershed or near vicinity of a lacustrine system include the following examples.

Example 7.6 On-site washing of gravels, placement of borrow, excavation, and dredging can result in the introduction of nutrients and/or toxic chemicals into the water column (see Ex. 7.4). The removal of bottom mud (through dredging) might also result in the removal of detritus essential for detritus food webs.

Regardless of their chemical nature, particles introduced into the water column will increase turbidity with possible effects on the compensation depth. While in suspension, these particles may cause abrasions on fish gills (depending on the abrasive index of the particles and on resident fish species) or (if clay sized) can coat fish gills, leading to suffocation. The particles may also settle out on developing fish eggs, resulting in the suffocation of fry.

Example 7.7 The erection of structures such as bridges can result in shadows being cast on areas that are significantly productive of vegetative biomass that is essential for the maintenance of grazing or detritus food webs.

Example 7.8 Rapid fertilization techniques used for slope stabilization and landscaping can lead to runoff of nutrients into the water column, with a consequent enhancement of eutrophication.

Example 7.9 Modification of the hydrological regime (e.g., through diversion of contributory streams, modification of spillways, placement of conduits) may alter the index of shoreline development. Increases in this index (i.e., increase in sinuous nature of shoreline) could result in increased habitat for shoreline vegetation and animal species that utilize that vegetation. Thus, such a change may also enhance the rate of eutrophication. Decreases in this index are likely to depress the rate of eutrophication.

Example 7.10 Structural maintenance activities, such as scraping and painting bridges or application of other types of preservatives to structures near water, can directly result in the introduction of toxic chemicals into the water column, with possible subsequent effects on food webs, including the bioaccumulation of toxics.

RIVERINE SYSTEMS

Key Dynamics

The basic ecological processes that take place in lacustrine systems also take place in riverine systems. However, in riverine systems, the flow of water essentially transforms time into distance. Thus, ecological processes are manifest along some length of the watercourse, the length being dependent on the rate of flow.

Figure 7.7 depicts the zonation of a river that typically results from a point source input of organic material. As shown in this figure, beginning at the point of discharge of the organic material and continuing some distance downstream, a *polysaprobic zone* (poly = many kinds of; saprobic = pertaining to decay) is established. In this zone, the color and turbidity of the water is relatively high because of the organic matter. Sunlight therefore cannot penetrate the water column to support photosynthesis. Because of the lack of photosynthesis (which produces oxygen) and the active growth of decomposing bacteria and fungi (which consume oxygen through respiration), the concentration of dissolved oxygen is very low; only anaerobic or microaerobic biota can survive. Finally, because of essentially anaerobic conditions, this zone is characterized by the foul smells typical of septic conditions.

As the introduced organic matter flows downstream, it is subject not only to decomposition but also to sedimentation. If the watercourse is long enough and if no additional inputs of organic matter occur, the breakup of organic particles (through decomposition) and their sedimentation will eventually result in a decrease in color and turbidity, with subsequent improvement in sunlight penetration of the water column. As shown in Figure 7.7, the final zone (*oligosaprobic zone;* oligo = sparse) is characterized by low color and turbidity and by the active photosynthesis of primary produc-

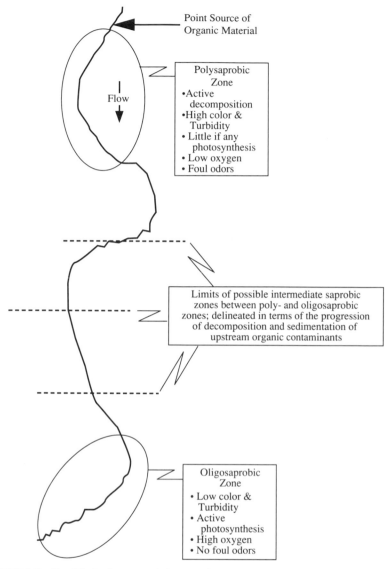

Point Source of
Organic Material

Flow

Polysaprobic
Zone
• Active
  decomposition
• High color &
  Turbidity
• Little if any
  photosynthesis
• Low oxygen
• Foul odors

Limits of possible intermediate saprobic
zones between poly- and oligosaprobic
zones; delineated in terms of the progression
of decomposition and sedimentation of
upstream organic contaminants

Oligosaprobic
Zone
• Low color &
  Turbidity
• Active
  photosynthesis
• High oxygen
• No foul odors

FIGURE 7.7 Simplified schematic of the saprobic zonation of a river induced by a point-source discharge of organic material.

ers, which utilize nutrients released by upstream decomposers and release oxygen into the water column.

The establishment and linkage of the various saprobic zones in a stream is often referred to as the *self-purification process* of that stream. As this phrase implies, running water will "purify" itself, as long as it has sufficient time (length of watercourse) to process organic inputs. However, if organic matter is released to the stream along its length, either through unregulated point source releases (e.g., sewage pipes) or enriched surface runoff (e.g., agricultural runoff), the polysaprobic zone will extend farther and farther downstream and even out into the ocean.

Another aspect of the ecology of lotic systems that should be given particular attention in impact assessment is stream bank vegetation. As in any ecological system, vegetation is not only a source of energy for herbivores (primary consumers), but also a source of energy-rich detritus. Lotic ecosystems depend not only on primary producers living within them but often also on stream bank vegetation that drops (or is periodically washed or scoured) into the water and thereby contributes its detritus to aquatic food webs. Externally produced detritus that enters a stream and supports its ecology is referred to as *allochthonous* detritus and, in many instances, can equal or exceed the amount produced within the stream (i.e., *autochthonous* detritus). Therefore, lotic ecology has often been described as "watershed ecology," a phrase that emphasizes the importance of surrounding land on the dynamics of lotic ecosystems.

Stream bank vegetation is also an important factor in the temperature regime of streams. The removal of overhanging vegetation can result in an increase in the temperature of surface water by as much as 10–15°C in some temperate areas. Generally, a temperature increase of 10°C over ambient conditions is considered significant with respect to any biological system. Moreover, as water heats up, its capacity to retain dissolved oxygen declines. For example, freshwater saturated with air at 10°C contains 10.92 ppm (mg/liter) dissolved oxygen, whereas at 30°C it contains a maximum of only 7.53 ppm dissolved oxygen. This difference could be particularly important in streams that have few or no riffle areas, which mechanically oxygenate water, and also receive a relatively high organic load—a situation that could result in persistently low concentrations of oxygen throughout extensive portions of the watercourse.

A key factor in lotic habitats is, of course, the velocity of water. Within any river, velocity can vary dramatically over different reaches of the river, depending on the surface gradient, the depth and width, and the geological nature of the river bed. Velocity can also vary with seasonal or other factors such as freshet flow (water inputs derived from snow melt), storms, downstream discharge from impoundments, and groundwater discharge. These spatial and temporal variations in velocity, in conjunction with spatial and temporal variations in oxygen, light, organic loading, and substrate underlie the richness in habitat diversity that can typically be associated with lotic systems.

Finally, lotic systems often play important roles in the migratory behavior of organisms. *Anadromous* fish, for example, include those species (e.g., salmon) that spend their adult life in salt water but migrate up rivers for spawning. *Catadromous* species include those species (e.g., freshwater eels) that spend their adult life in freshwater but migrate down rivers to the sea for spawning.

Examples of Direct and Indirect Impacts

Some impacts previously discussed with respect to lacustrine systems (Ex. 7.2, 7.6, 7.7, 7.8, 7.10) also apply to lotic systems. Additional possible impacts that directly or indirectly derive from project-related activities and the dynamics of lotic systems are presented here.

Example 7.11 Project design features that result in changes in the hydrological or hydrodynamic regime of a lotic system (e.g., placement of upstream dams, changes in controlled flow release from impoundments, removal of watershed vegetation, diversion of tributaries) may cause changes in any preexisting saprobic zonation, including an increase in the downstream extent of a polysaprobic zone. This condition could interfere with and even preclude certain uses of downstream reaches, including contact recreation, potable water supply, or commercial fisheries.

Example 7.12 In-stream or watershed excavation and/or land clearing might increase the particulate loading of streams and subsequent siltation of downstream habitats. Depending on the overall availability of these habitats, consequent impacts on biological species (e.g., benthic organisms) dependent on them may result in disruption of key food webs. If the particles are organic, subsequent decomposition may overwhelm the self-purification process of the stream, resulting in an extended polysaprobic zone.

Example 7.13 Project design features such as check dams, energy dissipators, and conduits as well as project-mediated changes in water quality parameters, such as temperature (due to removal of stream bank vegetation) and turbidity (due to excavation or land clearing), can act as barriers to the migration of anadromous and catadromous species.

The placement of some structures within the stream, including rock beds (to dissipate energy) and gabions (to stabilize a bank), may result in the introduction of new habitat into a river reach, with a consequent enhancement of species diversity or an increase in the population density of previously constrained species.

Example 7.14 The dredging of bottom muds (e.g., for the placement of bridge footings or piers, for improved navigation) may result in the loss of detritus not available elsewhere for downstream detritus food webs. The same mud may also serve as habitat not available elsewhere along the river for ecologically important species. The filling of riverine wetlands (see Chapter 9) associated with project development may also result in a significant loss of vegetation that serves as a primary source of detritus for riverine food webs.

Example 7.15 Because a lotic system is typically sensitive to the attributes of its watershed, projects far removed from a stream may nonetheless impact that stream.

- The atmospheric deposition of acids and other chemicals carried into a watershed by prevailing winds can result not only in the possible acidification of riverine waters, but also in their enrichment with cations (e.g., aluminum, zinc) leached out of watershed soils by acid rain;
- The runoff from highways and other bitumen surfaces can carry hydrocarbons and metallic dusts into the watershed and then into tributary surface and groundwater; and
- Long-term buildup of organic and inorganic nutrients and toxics in tributary surface and groundwater can be caused by human activities such as the disposal of sewage and other wastes, the application of fertilizers, and industrial development within riverine watersheds.

ESTUARINE SYSTEMS

Key Dynamics

As places where freshwater rivers empty freely into the sea, estuaries are characterized largely by their temporal and spatial variability in salinity. This variability is the result of the ocean's tidal action, the velocity of freshwater current flow into the estuary, wind velocity, and the detailed morphometry of the estuarine basin.

In some instances, the salinity of estuarine waters may vary significantly from surface to bottom, with less dense freshwater overlying the more saline and therefore more dense ocean water. However, sufficiently strong eddy currents caused by obstructed flow, tidal flow, or strong winds can overcome this chemical stratification and result in homogeneous salinity from surface to bottom. In some instances, again depending on the relative flow velocities of both fresh and saline water, saline water flowing upstream may override the seaward flowing freshwater, producing an inversion of density stratification. At some point in the estuary, however, density again becomes more important than current velocity and the layers mix.

The oscillating mixture of ocean water, having a salinity of about 35‰ (parts per thousand), and freshwater, having a salinity of 0.1–0.3‰, results in a linear pattern of increasing salinity from the mouth of the river to the ocean at high tide (Figure 7.8). The seaward flowing surface water in an estuary, which becomes increasingly saline and therefore heavier as it approaches the ocean, sinks into the counterflowing lens of salt water and is entrained in that lens. This entrainment results in the physical trapping of dissolved nutrients and suspended sediment in the upper (or landward) reaches of the estuary, as well as in replenishment of phytoplanktonic populations washed seaward by freshwater currents.

In the upper estuary, trapped nutrients are quickly taken up by phytoplankton, macroscopic algae, and rooted macrophytes (e.g., marsh grass species) that ultimately contribute their detritus back to seaward reaches of the estuary where many of the nutrients released through the mineralization of that detritus are trapped and circulated back to upper reaches.

Figure 7.9 summarizes some of the key roles that the mechanical energy of tide and waves plays in (1) recycling nutrients to primary producers and (2) circulating both primary producers and detritus to grazing and detritus food webs, many members of which are sessile (anchored or attached) organisms (e.g., oysters) that also depend on tides to flush away their metabolic wastes. In temperate regions, tides can also buffer seasonal temperature changes in estuarine waters; high summertime tides bring cooler water into the estuary, and low winter tides expose shallower areas so they warm more easily in the sun. This moderating influence on estuarine water temperature can promote an essentially constant rate of photosynthesis throughout the year.

Because of tidally influenced nutrient trapping and recycling, removal of metabolic wastes, circulation of detritus and primary producers to consumers, and moderation of temperature fluctuations, estuarine ecosystems are among the most biologically productive ecosystems in the world. Although much of this productivity, in terms of living biomass and detritus,

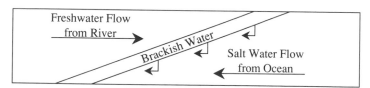

FIGURE 7.8 Cross-section of estuary. Heavier tidal water (from right) underlies less dense freshwater (flowing from left). Brackish water formed at the interface of fresh and salt water is more dense than freshwater and, along with its suspended particles and dissolved nutrients, becomes entrained by tidal flow up into the estuary.

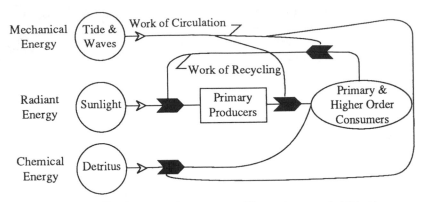

FIGURE 7.9 Energy flow in an estuarine ecosystem. The workgate symbol (black) represents a modulator (or regulator) of energy flow.

is recycled within the estuary, much also outwells to the sea where it can become an important energy input to offshore marine ecosystems.

Habitat constraints within an estuary are primarily defined by salinity (a physiological constraint) and current velocity (a physical constraint). Although many estuarine species are essentially marine species and are therefore tolerant of sea water salinity, some species are essentially restricted to estuarine habitats throughout their life cycles. Others are restricted to estuarine habitats only for certain stages of their life cycles (e.g., larval stages) and return to the sea in their adult stages. Much of the biomass of estuarine species is typically accounted for by benthic species that are securely attached to bottom substrate or otherwise protect themselves from displacement by currents by burrowing into bottom muds or attaching themselves to or among other anchored species or debris. Of course, estuarine species also include motile species, such as crustaceans and fish, as well as planktonic (freely floating) species. The distribution of motile species within the estuary is also largely dependent on their life stage and tolerance to salinity. Because the movement of plankton within an estuary depends on currents, their distribution is essentially determined by estuarine hydrodynamics.

Examples of Direct and Indirect Impacts

Some of the possible impacts previously discussed with respect to lacustrine and riverine systems (Ex 7.2, 7.7, 7.10, 7.14) also apply to estuarine systems. Examples of impacts that are most often associated with the unique dynamics and attributes of estuarine systems follow.

Example 7.16   Project design features that require the filling of intertidal marshes and mudflats are likely to result in a substantial reduction in the primary producers (e.g., marshgrass, diatoms) that act as the primary nutrient pumps in estuaries and provide the energy basis for the grazing and detritus food chains that characterize the estuarine ecosystem. A reduction in estuarine primary producers will also lead to important reductions in the outwelling of estuarine nutrients to the sea, and therefore have potential impacts on offshore productivity.

Filling these sensitive areas will also remove essential habitat and nutrients for marine species that use that habitat for the development of larval or juvenile forms. This change could result in significant reductions in certain marine populations, with possible subsequent economic effects on commercial fishery.

Finally, the removal of such habitat will reduce the capacity of the estuary to (1) provide habitat for migratory terrestrial species (e.g., migratory waterfowl species) and (2) mineralize organic material entrained in land runoff into the estuary (i.e., perform "tertiary waste treatment").

Example 7.17   Design features that might alter the hydrodynamics of an estuary (e.g., placement of fill; construction of bridge footings, piers, or jetties; dredging) could result in changes in the tidal influence within the estuary, with subsequent effects on (1) the distribution of species within the estuary, (2) nutrient trapping, and (3) the cycling of detritus and nutrients.

Similar consequences could also result from activities that affect the flow of freshwater into the estuary, including (1) channelization of tributary rivers and streams, (2) land clearing in the watershed, and (3) changes in land use that result in increased sheet flow from the watershed, such as that from parking lots and highways.

Example 7.18   Projects and activities not necessarily directly associated with estuaries (e.g., deposition of hazardous waste in the ocean, accidental maritime spills of oils and other chemicals, emptying and cleaning ship tanks, use of toxic paints and preservatives in marinas) may result in the tidal entrainment of toxic chemicals and, subsequently, an accumulation of those toxics in estuaries (because of their hydrodynamic trapping capacity). Because estuaries are typically nurseries for the larval stages of many commercially important marine species, the physical accumulation of toxics in estuaries could lead not only to bioaccumulation within estuarine food webs but also to bioaccumulation in humans.

Of course, because estuaries tend to be extensive and because they receive runoff or atmospheric depositions from extensive areas, other activities including the use of agricultural pesticides and the discharge of industrial wastes could also result in similar effects.

Example 7.19   Because much of the living biomass of an estuary is benthic, project-mediated deposition of sediment in estuaries becomes of special concern. Extensive land clearing in areas with high rainfall (e.g., Southeast Asia, South America) may result in such massive sedimentation in a brief period of time that it will overcome not only any scouring effect of tides but also the biological capacity of benthic species, especially sessile types, to survive engulfment. Very high sediment loading, should it occur during critical migratory periods, could also seriously impact anadromous and catadromous fish species. Should it occur during critical reproductive or developmental periods for other aquatic species, sediment loading could also impact those species by effectively removing nesting habitat, covering developing fry, and/or abrading gills.

GROUNDWATER RESOURCES

Key Dynamics

Rainfall percolating down through the soil travels through the interstitial spaces among soil particles. As long as this water encounters interstitial spaces that are filled with air, the rainwater replaces the air and, consequently, continues to percolate downward. At some depth (called the *water table*), all the interstitial spaces become filled with water and the movement of water is directed from the vertical to the horizontal direction. As shown in Figure 7.10, that layer of soil in which all interstitial spaces are filled with water, that is bound at the surface by the water table and at the bottom by an impervious layer of soil (e.g., bedrock, densely packed clay), is the *water table aquifer*. This region is, in effect, a soil pipe that horizontally transfers water from one place to another. In many instances, a water table aquifer may be a significant source of water to impoundments and rivers.

Wells that derive their water from water table aquifers are appropriately called water table wells. Such wells may be relatively shallow or deep, depending on the depth of the water table as measured from the surface of land at the point where the well is constructed. Whatever their depth, water table wells share one important characteristic. The quality of water in water table wells is directly influenced by the quality of water that percolates down from the surface to the water table.

The surface area of land that overlies a water table aquifer and through which surface water must percolate to enter that aquifer is known as the *recharge area* for that aquifer. As indicated in Figure 7.10, the recharge area for a water table aquifer is not only extensive but is also inclusive of the immediate area in which the well is located. Uncontrolled land use in the extensive recharge area (e.g., use of agricultural pesticides, industrial

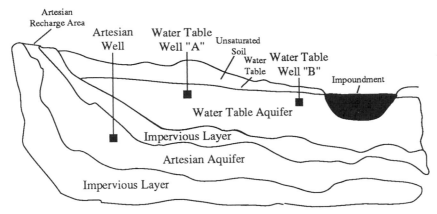

FIGURE 7.10   Cross-section of water table and artesian aquifer. Water table wells (A and B) may be of different depths, depending on surface topography and depth of water table. Artesian wells tap into a deeper artesian aquifer that is bound by impervious geological strata.

discharge, highway runoff) as well as activities conducted in the immediate area of the well (e.g., household use of pesticides, land disposal of household sewage) can lead to immediate contamination of the water table aquifer.

Another type of aquifer is the *artesian aquifer*. In this case, aquifer water is constrained at the top and the bottom by an impervious layer. An artesian aquifer also has its recharge area, but this area is typically far removed from the place at which water is drawn out through an artesian well. The recharge area may be 10, 100, 1000, or more miles away. Thus, the time required for surface water to flow through an artesian aquifer and into an artesian well can be substantially greater than the time required for surface water to flow through a water table aquifer and into a water table well. For example, a measured flow through an artesian system may take 100 years whereas only 10 days are required in a groundwater system. This difference is important with respect to the contamination of recharge areas with sewage waste, for which the primary concern is human pathogens. Human pathogens released into the environment from the gastrointestinal tract of humans die off in several weeks. Thus, contamination of water table aquifers with sewage often results in the spreading of human disease because flow rates are relatively fast. The pathogens do not have enough time to die. This problem does not typically arise for artesian aquifers. Note that no difference in the degree of hazard or risk exists between the chemical rather than the biological contamination of artesian recharge areas and water table recharge areas. Many toxic chemicals are highly soluble (and are therefore not filtered out by passing through soil) and highly persistent (i.e., they are not easily biodegraded in soils).

Because an artesian aquifer is constrained on both sides by an impervious layer and because that layer must be slanted up to the surface of the land somewhere (through its recharge area), water in artesian aquifers can be subject to significant hydraulic pressure. In some instances, the pressure is sufficient for water to rise to the surface of an artesian well. However, the pressure is often insufficient, and submersible pumps must be used to draw the water.

Water contained within an artesian aquifer may mix with water contained within a water table aquifer if the impervious stratum separating the two is fractured or contains fissures. This mixing can also occur if the well casing of an artesian well is not properly sealed against the bore hole that is drilled through the impervious layer.

Figure 7.11 depicts the cone of depression in the water table. This phenomenon occurs during the operation of a well pump, and represents the drawdown of the water table, which is maximum in the immediate vicinity of the well and decreases with distance from the well. As also depicted in the figure, two or more wells may be located so their cones of depression overlap, resulting in a possibly significant localized drawdown of the water table. A sudden and significant drawdown of the water table can, of course, have negative consequences for vegetation if that vegetation is dependent on the water table.

Examples of Direct and Indirect Impacts

Example 7.20 Blasting during construction may result in local fracturing of the impervious boundary of an artesian aquifer, leading to the flow of water from a water table aquifer into an artesian aquifer and then into an artesian well. If the water table aquifer is contaminated with pathogens (sewage waste) or chemicals, its mixture with artesian waters may pose an immediate public health hazard to individuals using the artesian aquifer as a potable water supply.

Blast-induced fracturing of impervious substrata may also result in a lowering of the water table because of the deeper percolation of groundwater through the fractures. Should this occur, water table wells may become nonfunctional and vegetation (including agricultural crops) may wilt.

Example 7.21 Excavation and landfilling, as well as the placing of subsurface foundations, may result in redirected groundwater flow, including the flow of leachate from a septic system to a previously protected water table well.

Example 7.22 Well field development may result in a persistent drawdown of the water table, leading to the loss of vegetation and subse-

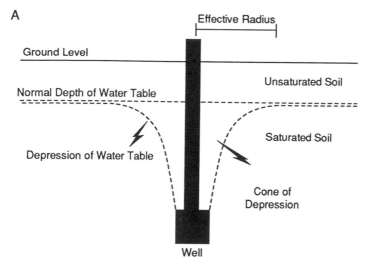

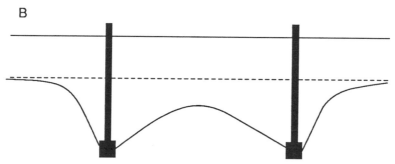

FIGURE 7.11 Cross-section of geological stratum, showing cones of depression of two nearby water table wells. (A) When well pumps, water table is depressed in immediate area of well. The degree of depression is greater for greater pumping rate. The linear extent of depression from the well is the "effective radius." (B) When two or more wells in the near vicinity pump, cones of depression can combine, resulting in significant drawdown of water table in area of overlapping radii of influence.

quent soil erosion. Uncontrolled well field development of an artesian aquifer can result in a localized drawdown of the artesian aquifer, leading to localized well failures.

Example 7.23 Application of pesticides (during landscaping and maintenance phases of project development), on-site processing of construction materials (gravel washing), the placement of borrow that has not been analyzed for possible chemical contamination, and on-site demolition of structures possibly containing hazardous materials (e.g., asbestos) can result in the contamination of aquifer recharge areas (for artesian and water table aquifers) and subsequent public health consequences.

Example 7.24   The use of heavy equipment in an aquifer recharge area may result in sufficient soil compaction to increase surficial runoff and decrease percolation rates, thus significantly reducing recharge of the aquifer. Similarly, the establishment of extensive paving within a recharge area and the placement of culverts and drainage ditches to direct flow away from a recharge area will also reduce recharge and is likely to result in the failure of existing wells. Reduced recharge could also lead to reduced flow in riverine systems that may depend on groundwater contributions and also may constitute the primary input of freshwater into downstream estuarine ecosystems.

CHAPTER 8

TERRESTRIAL RESOURCES

Terrestrial resources, like aquatic resources, comprise interrelated abiotic and biotic components. As in aquatic environments, material and energetic transformations in the terrestrial environment are sequentially linked by grazing and detritus food webs that accomplish the ecological work of primary production, consumption, and decomposition. As shown in Figure 8.1, primary producers include the trees, shrubs, forbs (wildflowers), and grasses; primary consumers include a large diversity of mammals, insects, birds, reptiles, and amphibians; secondary, tertiary, and higher order consumers include many different types of predators; and, finally, decomposers include bacteria, fungi, earthworms, and numerous other biota that mineralize the organic molecules of detritus and release essential plant nutrients back to the soil.

Differences in the biology of terrestrial and aquatic environments are directly related to differences in habitat constraints in these two types of environment. General constraints include those related to (1) gravity, (2) oxygen concentration, (3) temperature, (4) availability of water, and (5) the nature of available substrate. For example, water is itself a supporting medium and imparts a buoyancy to living things that counteracts the effects of gravity. However, in the terrestrial environment essentially no counterforce to gravity exists. Terrestrial biota must, therefore, be adapted to withstand gravitational stress. Such adaptation is manifest in special supportive structures, such as root systems, the woody tissue of trees and shrubs, and the relatively massive bone tissue of vertebrates.

Water also acts as a buffer against abrupt changes in temperature; however, air does not. Thus, the daily temperature of the terrestrial environment may (depending on specific location) vary greatly in relatively short periods of time. Terrestrial biota living within areas with such variations in temperature must, accordingly, be adapted to these changes in terms of their behavior, their physical attributes, and their physiological capacities. These

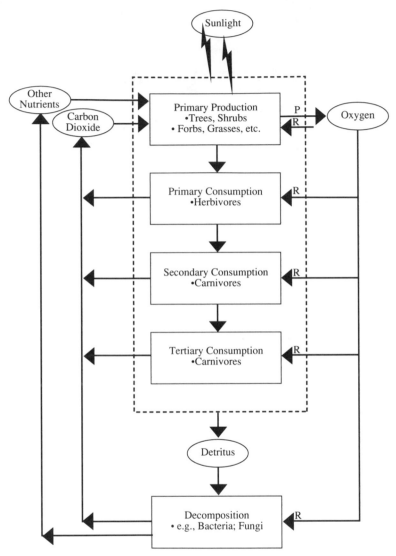

FIGURE 8.1 Simplified schematic of interconnected niche functions in a terrestrial ecosystem. Arrows represent the flow of matter, which is recycled by systemic dynamics. P represents the photosynthetic production of oxygen; R represents the respiratory consumption of oxygen.

and other examples of very broad types of biological adaptations to the constraints of terrestrial and aquatic habitats are summarized in Table. 8.1.

HABITATS

As "places" used by organisms, habitats should be viewed as having vertical, horizontal, and temporal constraints. In other words, geographical and temporal constraints or limits essentially define the utility of a particular "place" to a particular organism, population, or community. For purposes of identifying possible impacts on terrestrial ecosystems, the assessment team must

1. identify specific habitat constraints within the proposed project area,

TABLE 8.1 Generalized Biological Adaptations to the Constraints of Terrestrial and Aquatic Habitats

Habitat constraint	Comments on terrestrial environment and biota	Comments on aquatic environment and biota
Gravity	Exo- and endoskeletons to support biomass; supportive structures can be substantial percentage of total biomass of organism	Water itself acts to support biomass; adaptations primarily to maximize or utilize buoyancy
Oxygen concentration	Concentration in air is generally the same over space and time; can vary greatly in soils, depending on water content; some specific plant adaptations to wet soils	Can vary greatly with location and time; aquatic biota often adapted to spatial and temporal variations
Temperature variation	Can vary greatly with location and time; terrestrial biota often adapted to spatial and temporal variation	Heat capacity of water tends to prevent abrupt change; aquatic biota generally not adapted to abrupt or large changes
Availability of water	Can vary greatly with location and time; terrestrial biota often adapted to spatial and temporal variation	Generally not a constraint to aquatic biota
Nature of substrate	Generally stable and therefore not usually a major constraint to terrestrial biota	Generally unstable and transitory; many types of adaptation to temporal changes in substrate

2. relate geographic and temporal variations of these constraints to the behavioral, structural, and physiological adaptations of site-specific floral and faunal populations,

3. relate project-mediated impacts on habitat constraints to consequent effects on these populations, and

4. evaluate the overall consequences of these impacts on the ecological dynamics of terrestrial communities.

Identifying Habitat Constraints

Two basic field-study approaches may be used to identify important geographic and temporal aspects of habitat—a reconnaissance-type study and a vegetative analysis.

A *reconnaissance-type study* of the project area and its environs is typically used for the evaluation of a large-area habitat and includes various techniques for evaluating the suitability of the study area for individual animal species. Such techniques include:

1. **Cover mapping:** Mapping of an area with symbols that represent vegetative cover types, including overstory (e.g., trees), understory (e.g., shrubs), and ground cover (e.g., mosses, grasses)

2. **Cover density mapping:** Mapping of an area with symbols that represent qualitative and quantitative aspects of the amount or degree of cover by different vegetative types

3. **Soil mapping:** Mapping an area with respect to soil types, including physical and chemical characteristics of that soil (e.g., soil moisture, temperature, geological derivation, nutrient content, areas of known chemical contamination, erosion-prone soils)

4. **Food mapping:** Mapping of an area with respect to the total yearly crop (i.e., of food required for specific species), the quantity of food available at a particular time (e.g., seasonal maxima and minima), and the proportion of the available food that is actually consumed

5. **Indicator species mapping:** Mapping an area in terms of individual plant species that indicate specific conditions of climate, soil type and moisture, and previous disturbance

A *vegetative analysis* of habitat in the proposed project area and its environs is typically used for a detailed examination of the vegetative aspects of wildlife habitat. A variety of techniques may be employed to measure one or more of the following:

1. **Species presence/absence:** The simple presence or absence of a certain species (always necessary information when assessing areas

that contain the habitat of legally protected plant or animal species)

2. **Basal area:** The proportion of ground surface occupied by a particular species
3. **Cover:** The vertical projection of above-ground (overstory) parts onto the ground
4. **Frequency:** The percentage of sample plots in which a particular species occurs
5. **Density:** The number of individuals in a species per unit area
6. **Dominance:** Some estimate of the comparative population size of different species
7. **Importance:** The value assigned to a given species with respect to a defined goal or objective (e.g., primary food supply for a certain animal species)

With respect to these basic approaches to evaluating habitat constraints, field studies undertaken for the purpose of habitat evaluation require professional inputs in the design, operational, and analytical phases of such studies. Thus, from the practical perspective of impact assessment, which is often severely constrained by a limited budget and the availability of specialized personnel, the assessment team should first assume that some literature (including reconnaissance-type and vegetative habitat analysis) does exist and is relevant to the general project area under consideration. Although this assumption may ultimately prove to be incorrect, the assessment team will at least have first exhausted possible sources of information before committing itself to expensive and time-consuming field studies.

Appropriate governmental agencies and organizations that should be contacted for possible information include fish and wildlife agencies (at local, regional, and national levels), geological survey agencies, environmental quality and natural resource agencies, local and regional universities, and local and regional wildlife or conservation organizations. Consulting other environmental impact studies (EISs) that may have been completed for previous or ongoing projects in the general project area is also important. Such EISs may have directly relevant information or may identify additional sources of information. The assessment team should also be aware that, although such agencies and organizations may not have actual habitat maps of the general project area, they may have tabulations of soil and vegetation types, aerial photographs (black and white, color, color-infrared), and remote-sensing data that can be used and transferred to map overlays.

Two key concepts are useful for soliciting appropriate data and information from agencies and organizations—plant associations and biomes. A *plant association* is a collection of plant species that have a high probability of occurring together in a local area. A plant association may include relatively few species, or it may include dozens of species. Individual plant associations (also called plant formations or vegetative types) are typically named after the predominant species in the association, or after two or more prevalent species. In some instances, plant associations may also be assigned a code number as, for example, in the Society of American Foresters guide to tree associations in North America.

Examples of the types of information to be sought regarding plant associations include:

- the kinds of plant associations present in the project area and its environs,
- the number of acres or proportion of total area covered by different plant associations,
- the individual species included in identified plant associations that are most likely to occur in the project area and its environs, and
- the relative abundance of identified plant associations in the general region, that is, whether they are common or uncommon

A *biome* is a community of plant and animal species that is characteristic of a general climatic area. Biomes are often named after the dominant type of vegetation, for example, coniferous forest biome, deciduous forest biome, or grassland biome. However, the concept of biome is a broad community-based concept that includes consideration of the interrelationships of plants and animals. Thus, in soliciting information on the biome(s) in which a proposed project is to be located, the assessment team is soliciting not only information on relevant plant associations but also information on the types of animal species typically found in habitats characterized by the presence of those plant associations. Also, much of the published literature generally available on biomes contains specific information on the types of parameters or physical and chemical attributes particularly important for the maintenance of specific biomes (e.g., the role of deep-rooted trees located at the periphery of a deciduous forest as wind buffers to more shallowly rooted trees located within the forest).

In many instances, the assessment team will find little specific information on the habitats of the site-specific project area. In such cases, the assessment team should:

- compile whatever information is available on soils, climate, hydrology, vegetation, and so on;
- consult local botanists, zoologists, and ecologists (and/or the general literature) to identify possible or likely plant associations and animal populations in the project area and its environs; and
- only when no information is available that can be reasonably extrapolated to site-specific conditions, conduct the aerial and field reconnaissance that may be required to identify the type, location, and extent of plant and animal association.

On the basis of more than 20 years of experience with impact assessment, we consider it highly improbable that the assessment team will have to undertake detailed, long-term field studies of habitat to complete a reasonable assessment of project impacts. The world literature base on biomes is replete with data and information on the various types of habitats typically available within the various types of biomes. Much of this information can be reasonably extrapolated to site-specific conditions, especially if general field reconnaissance is performed to "ground-truth" (i.e., use field data to verify) general expectations that derive from such extrapolations. However, extensive and possibly long-term field studies may be required to evaluate fully project impacts on legally protected species (e.g., threatened or endangered species). In such instances, professional wildlife biologists may need to design and undertake the study and evaluate the generated data.

Relating Habitat to Biological Adaptations

For purposes of impact assessment, plants and animals found within the project area and its environs are best assumed to be specifically adapted to their habitats in terms of their structure, physiology, and behavior. Therefore, the assessment team should become familiar with some practical concepts that are useful for understanding biological adaptations. Such concepts can greatly facilitate the information-gathering phase of impact assessment by focusing the team's attention on key issues to be considered.

An example of an important concept that gives perspective to the whole question of biological adaptation is *Shelford's law of tolerance*. This law (Figure 8.2) states that an optimum condition exists for any species regarding any physical or chemical parameter of its habitat; a situation representing less than (e.g., too low a temperature) or more than (e.g., too high a temperature) the optimum is, therefore, to the detriment of that species. Two points should be emphasized with respect to Shelford's law and its depiction in Figure 8.2

1. A common way in which to express this law is to say that any species (plant or animal) possesses a certain **tolerance** to any variable aspect of habitat. Its tolerance may be relatively narrow or relatively broad, but should the conditions of habitat exceed the lower or the higher limit, the habitat becomes inimical to that species. In this sense, then, a species is **adapted to specific ranges of variability among the parameters that define its habitat.** When the species' structural, physiological, and/or behavioral adaptations become insufficient to deal with actual variations in habitat (mediated, perhaps, by project-related activities and conditions), the species is likely to disappear from that habitat or at least to suffer by remaining within it.

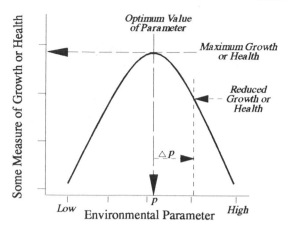

FIGURE 8.2 Biological response to variation in an environmental parameter. There is always an optimum value (*p*) of any environmental parameter for the selected measure of response.

2. That the disappearance of a species from its habitat will result in an "empty" habitat is highly improbable. The disappearance of one species is, in fact, the usual first step in a sequential colonization of the altered habitat by another species specifically tolerant to the conditions of that altered habitat.

In light of these considerations, the assessment team should understand that project activities that can affect successful adaptation by existing populations have the potential not only for causing the demise of those particular populations, but also for initiating a biological succession in the affected area. Frequently, assessment teams have focused only on populations that might be excluded from the project area and have given little if any attention to those populations that may be introduced. In any comprehensive assessment, both the exclusions and the introductions of populations that might be mediated by project development must be considered.

Whereas Shelford's rule has general applicability to all types of ecosystem, a concept with more limited application to the terrestrial environment is the *root-to-shoot ratio*. Although the root-to-shoot ratio is a numerical ratio between below-ground biomass (root) and above-ground biomass (shoot) of vegetation, we refer to it as a concept because the numerical quantity is as much a tool for understanding complex and interrelated phenomena as it is for measuring a particular attribute of vegetation.

A rooted terrestrial plant is a living machine that does the ecological work of primary production. However, these plants are essentially two interconnected machines. The above-ground component carries on photosynthesis and, therefore, requires water and nutrients that are supplied by the below-ground component. The below-ground component, in turn, requires food, in the form of energy, that is supplied by the above-ground component and is metabolized in the below-ground component to extract water and dissolved nutrients from the surfaces of soil particles to which they are tightly bound by surface tension. If the roots do not receive energy from photosynthesizing leaves, water and nutrients will not be obtained. Similarly, if water and nutrients are not obtained, the leaves will not be able to carry on photosynthesis. The root-to-shoot ratio is, therefore, a numerical representation of the particular balance reached between a plant's energy-producing leaves and water- and nutrient-supplying roots. This balance may be assumed to be optimum for a particular plant in a particular habitat, but may be influenced by project-mediated changes in that habitat.

For example, alteration of soil percolation rates (by soil compaction due to the operation of heavy equipment) or covering leaves with light-occluding dusts (resulting from land clearing operations) can result in a re-adjustment of an exisiting root-to-shoot ratio. One consequence of such a re-adjustment may be, of course, wilting of the plant. If a new and adequate balance of roots and shoots cannot be achieved, the plant will die.

Another concept that is important in relating terrestrial habitats in project areas and their environs to biological adaptation is the general biological concept of *population dispersal*. Population dispersal, which should be seen as behavioral adaptations to habitat constraints, includes the following phenomena:

1. **Emigration:** One-way movement of a population out of an area
2. **Immigration:** One-way movement of a population into an area
3. **Migration:** Periodic departure and return of a population; includes
   - nocturnal migration—occurring at night
   - diurnal migration—occurring during daylight
   - diel migration—occurring over 24 hours
   - seasonal migration—occurring over seasons

The spatial area that a particular population (i.e., groups of individuals of the same species) utilizes to maintain itself, which is inclusive of areas defined by the migrations of that population, is called the population's *home range*. Projects and project-related activities can interfere with each of these behavioral adaptations to an organism's habitat. For example, security lighting may affect nocturnal migration. Equipment operation during the day, which produces loud noise, may affect diurnal migration. Fencing may affect diel migration. A range of design and structural features of a project (e.g., filling of small impoundments used by migratory birds, location of highways and pipelines within migratory pathway of large mammals) may interfere with normal seasonal migration and effectively reduce a population's access to its home range.

Impacts on Habitat Constraints and Biological Adaptations

Two basic steps are required in identifying possible impacts on habitat constraints and biological adaptations:

- identify the actual uses of existing vegetation by other terrestrial biota and
- identify project-mediated impacts on existing vegetation and on the use of this vegetation by terrestrial biota.

Plant associations and species perform various functions for local and migratory biological populations, providing

- food (both living vegetation and its detritus)
- shelter (short- and long-term)
- sites for breeding and rearing of offspring
- materials for nest-building

A key concept that integrates these functions, and is particularly useful for impact assessment, is *carrying capacity*. The carrying capacity of an area is the potential of that area to support an animal population. Carrying capacity is typically expressed as a number (i.e., the number of organisms of a particular species per unit area of land). Although the determination of carrying capacity requires the professional judgment of wildlife biologists, the assessment team can generally collect important information on carrying capacity in the project area without recourse to extensive field studies by

- estimating acreage in project area for each type of plant association;
- estimating acreage of each plant association to be clear-cut during construction;
- estimating acreage to be replanted with a particular type of association;
- using existing literature and/or local expertise to develop a list of individual species in each association;
- using information on existing associations and individual plant species, and on proposed plantations, to identify animal species likely to depend on those plant species; and
- consulting the published literature, appropriate governmental agencies, and wildlife biologists with local experience to gather estimates of the current (i.e., pre-project) and projected (i.e., after project implementation) carrying capacity of the area for selected species (Figure 8.3).

Assessors should not assume that projects always result in a decrease in carrying capacity for a particular species. In fact, projects can often increase the carrying capacity of the project area for populations that are

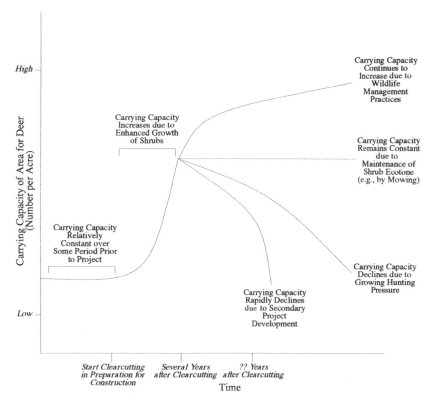

FIGURE 8.3 Some possible changes in a forest's carrying capacity for deer as a result of project development.

already present but are constrained by the availability of food. In using the concept of carrying capacity, the assessment team should carefully consider the following caveats:

1. Carrying capacity is essentially a judgment expressed as a number. Usually the number is expressed as a range which is often relatively large (e.g., 5–30 white-tailed deer per acre). The range may express the range of opinion of wildlife biologists or the range of conditions considered and differentially weighted. In comparing pre- and post-project estimates, establishing documented criteria (e.g., ± 10%) for determining significant differences is therefore important.

2. That estimates of carrying capacity can be obtained for any but a relatively small number of species that occur within the project area and its environs—these species being of particular interest for a variety reasons, including their economic and recreational potential and their legal status as protected species—is extremely improbable.

3. Assuming that the carrying capacity of any terrestrial environment for any particular species is at its maximum has become standard practice in environmental impact assessment. Thus, project-mediated decreases in the available habitat for a particular species are considered to result in the absolute loss of populations utilizing that habitat, that is, the assumption of maximum carrying capacity means that populations displaced from one location by a project cannot move to a new location because that new location is also at its maximum carrying capacity.

Project-mediated impacts on migratory behavior, root-to-shoot ratios, and carrying capacity represent only a few of the possible impacts on habitats and biological adaptations. As a general guide, the assessment team should consider that any project-mediated change in any on-going process of energetic and/or material introduction, translocation, dissipation, concentration, transformation, and elimination (Chapter 2) into, within, and from terrestrial components may result in impacts on the terrestrial habitat and on the successful use of biological adaptations by terrestrial biota.

KEY ECOLOGICAL DYNAMICS

In addition to project-mediated impacts (direct and indirect) on the habitats that terrestrial species utilize and on the successful use of biological adaptations by those species, project activities and attributes may also result in changes in dynamic ecological processes that support and depend on the structure and function of terrestrial communities. Two such processes are *ecological succession* and *nutrient cycling*.

Ecological Succession

Succession refers to predictable ecological changes. An example of succession in the aquatic environment is the eutrophication of lakes (Chapter 7). Examples of succession in the terrestrial environment include *primary* and *secondary succession*. Primary and secondary succession (also called ecological succession) are characterized by predictable changes in species and in community processes that (1) result from modifications of the physical environment caused by the biota that use that environment and (2) culminate in an ecosystem that is stable with respect to its overall climate. Primary succession begins with bare ground; secondary succession begins in areas that previously have been vegetated. The evaluation of ecological impacts of most projects typically includes consideration of impacts on secondary succession.

The sequence of biological communities that replace one another during secondary succession is referred to as the *sere*. The sere may be subdivided into three general *seral stages:* the pioneer stage, developmental stages, and the climax stage (Figure 8.4).

In the pioneer stage, plant and animal species may be generally characterized as opportunistic. They are typically small, have short and simple life histories, and reproduce rapidly. The community as a whole tends to be characterized by simple grazing food webs, rates of primary production that are greater than rates of respiratory consumption, and poorly organized spatial patterns of vertical and horizontal distribution.

In the climax stage, organisms are generally large, have long and complex life cycles, and are less apt to reproduce rapidly. The community as a whole tends to be characterized by detritus food webs, rates of primary

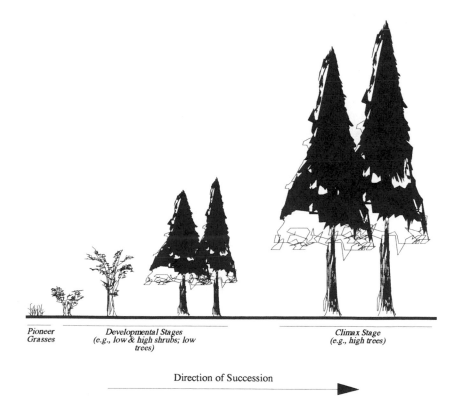

| Pioneer | Developmental Stages | Climax Stage |
| Grasses | (e.g., low & high shrubs; low trees) | (e.g., high trees) |

Direction of Succession

FIGURE 8.4 Example of plant and ecological succession leading from a grassland to a high tree forest. As plant associations change from one seral stage to another, so do the animal populations that utilize plant species for food and shelter. Some animal populations may be present throughout the succession, while others can utilize only certain seral stages.

production that are roughly equal to rates of respiratory consumption, and highly organized zones of vertical and horizontal patterns of distribution.

Organismic and community characteristics of developmental stages are intermediate to the characteristics of pioneer and climax stages.

As a general rule, the best information on the stage of succession in a proposed project area is available from local sources, including regional offices of governmental agencies, local colleges and universities, and local botanical and wildlife associations. Contacting these or other appropriate organizations to determine (1) the present seral stage in the project area, (2) probable future seral stages if the proposed project is not undertaken, and (3) probable future seral stages if the proposed project is undertaken is important.

The importance of these determinations lies in the fact that ecological succession is a natural process and, even in the absence of human activity, results in the sequential replacement of one biological community by another until some steady state between the community and its abiotic environment is reached (i.e., when the climax stage is established). Thus, isolating project-mediated changes in populations from natural successional changes in populations is necessary. For example, as plant succession proceeds from pioneer stages to developmental stages, habitat changes will result in the loss of certain animal populations. A particular project may speed up the loss of such animals by removing habitat more quickly that it would otherwise disappear through succession. A project may also prevent the loss of such animals by keeping the project area in an early phase of succession.

That a typical project will alter the ecological succession in an entire region is highly improbable. Typically, projects impact highly localized successions, usually resulting in the establishment of an *ecotone* (or edge), a zone of transition between different types of communities (Figure 8.5). Examples of ecotones include the zone of shrub growth between a forest and an open field, and the demarcation line between tall native grasses and short hybridized grasses.

Some types of ecotone are often viewed as desirable consequences of project development. For example, a shrubby ecotone between a mature forest and mowed-grass embankments along a highway can be an important food supply for certain herbivores, such as deer. However, the ecotone may effectively draw deer too near the highway and thus result in an increase in roadkill. Also so many ecotones can be created in a forest, for example, that the forest becomes an open field. Given the typically large range of consequences associated with ecotones, the assessment team must evaluate the overall consequences of ecotones with respect to regional ecology.

The importance of a regional overview in assessing local impacts on terrestrial ecosystems cannot be overemphasized. Professional ecologists tend to deal with whole forests, not simply a few acres, and with whole grasslands, not simply a few square miles. In other words, they tend to

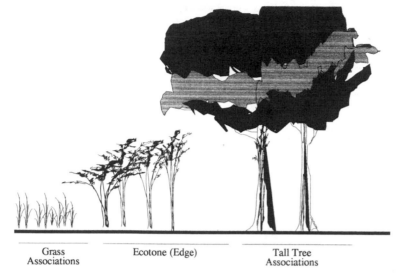

Grass Associations	Ecotone (Edge)	Tall Tree Associations

FIGURE 8.5 Example of an ecotone that separates plant associations to either side.

take a biome perspective. Project impacts along an alignment, for example, through a forest, should be viewed in terms of the forest ecology of the overall region and in terms of the interrelationships between the forest and other biomes in that region.

How large, then, is a region? No general rule exists that can be used by the assessment team to delineate the acreage of concern. The size of the region depends on particular interrelationships among specific primary producers, consumers, and decomposers. However wide the area required to identify basic interrelationships among these living energy transformers is precisely the area required for a comprehensive understanding of potential impacts on the terrestrial ecosystem characterized by those transformers. Whatever the size of that area might be, it is quite certain to extend beyond the property lines and rights-of-way of the proposed project.

Nutrient Cycling

Rooted terrestrial vegetation plays an essential role in nutrient cycling because it functions as a nutrient pump (Figure 8.6), raising dissolved nutrients out of the soil and, after transforming these relatively simple mineral nutrients into more complex organic molecules (through photosynthesis), releasing detritus back to the soil. The subsequent mineralization of detritus by such decomposers as soil bacteria and fungi completes the cycling of nutrients back to the soil. The deeper the penetration of roots into the soil,

Basic Nutrient Cycling Process

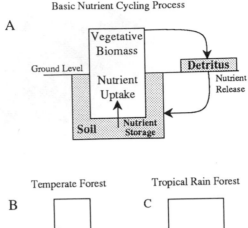

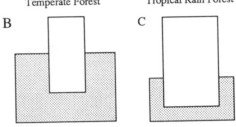

FIGURE 8.6 The basic nutrient cycling process (A) involves the vegetative uptake of nutrients from soil and subsequent release of those nutrients back to the soil through the decomposition of detritus. In a temperate forest (B), seasonal cold and dry periods result in a relatively slow rate of decomposition, resulting in the larger portion of nutrients being sequestered in the soil. In a tropical rain forest (C), the rate of decomposition and nutrient uptake is very rapid, resulting in the larger portion of nutrients being sequestered in the vegetative biomass.

the deeper the nutrient cycling process extends. Also, the more rapid the rate of vegetative growth, the more rapid the nutrient cycling process becomes.

In temperate forests, in which the rate of vegetative growth (including the growth of mineralizing soil bacteria and fungi as well as trees) is constrained by seasonal cold and dry periods, the rate of nutrient cycling is typically highly variable. In tropical and subtropical climates, where temperature and the availability of water are essentially optimal for vegetative growth throughout the year, the rate of nutrient cycling is relatively high and constant. Such differences in the rates of nutrient cycling in temperate and tropical or subtropical forests are critical factors in the ecological dynamics of these forest ecosystems.

For example, in tropical and subtropical forests, the larger proportion of available nutrients typically tends to be incorporated in the organic molecules that constitute the biomass of trees; the smaller proportion of available nutrients is incorporated into the soil because high ambient heat and moisture promotes the rapid mineralization of detritus and the rapid extraction of mineralized nutrients from the soil and their incorporation into the biomass of trees. In temperate forests, however, the reverse occurs. Slower rates of mineralization (contributions of inorganic nutrients to soil) and slower rates of vegetative growth (dependent on the extraction of inorganic nutrients from the soil) result in the accumulation of large amounts of organic materials in the soil, that is, the buildup of soil humus, which acts as a large reservoir of nutrients.

EXAMPLES OF DIRECT AND INDIRECT IMPACTS

Example 8.1   Clear-cutting of forested areas in tropical and subtropical forests will typically result in the loss of a significant reservoir of nutrients (i.e., nutrients incorporated into tree biomass). This reduction in available plant nutrients could result in greatly diminished secondary growth which, coupled with the original removal of forest canopy, would result in greatly enhanced erosion of exposed forest soils. Such eroded soils can, in turn, result in a variety of direct and indirect impacts on downstream receiving waters (Ex. 7.6, 7.12).

Example 8.2   Clear-cutting of any forested area may result in the exposure of previously protected trees to the full force of prevailing winds. Generally, such trees (i.e., those growing in the inner portions of a forest) have a less developed root structure and, when exposed to winds, become particularly vulnerable to being toppled. Under some circumstances, even highly localized clear-cutting may result in a progressively expanding zone of toppled trees, with consequent potential loss of forest habitat, carrying capacity, and even soil.

The introduction of winds into forested areas by clear-cutting is important not only because of their potential to topple shallow-rooted trees, but also because introduced winds may enhance the rate of evaporation of water from forest soils and change the relative humidity in understory habitats. Such changes in surficial hydrology and microclimate could mediate subsequent effects on particularly sensitive understory and on wildlife species dependent on that understory.

Example 8.3   The use of area lighting (e.g., tower lights) may be associated with a variety of potential impacts on vegetation. For example, depending on the nature of the lamps (e.g., mercury or sodium vapor), tower lights may adversely affect the growth of agricultural crops. They may also extend the growing season of certain temperate vegetative species which, in periods of frost, may result in extensive vegetative loss with consequent impact on both detritus and grazing food webs.

Example 8.4   The placement of fill materials can result in changes in local successional seres. This event typically occurs when fill material significantly differs from site-specific soils in terms of chemical or physical properties. For example, mud-flat soils (e.g., dredged from a tidal marsh) used as fill in a highland area can be expected to result in seres significantly different from naturally occurring seres in that highland area. In some instances (and based on local experience), this effect may be purposely used to accomplish landscaping objectives. Changes in natural succession may also be mediated by hydrological changes resulting from soil compaction (e.g., use of heavy equipment) and any alteration of topographical features that significantly affects surficial runoff.

To minimize impacts on local successional patterns, local soils should be used as fill material, with particular attention given to layering the fill in a manner that duplicates the various layers (or soil horizons) of undisturbed local soil. Soils along heavy equipment access roads and within staging areas should be returned to their previous natural condition. Surficial runoff should be managed to maintain local groundwater levels and minimize erosion.

Example 8.5   As in aquatic environments (e.g., Ex. 7.2), the potential biomagnification (including trophic magnification) of toxic chemicals within terrestrial food webs may be associated with a variety of project-related activities, including the use of pesticides along rights-of-way during operational and maintenance phases of project development; the pre-construction demolition of in-place structures that may contain hazardous materials; the use of fill material that may contain chemical contaminants; and the on-site storage or processing of construction materials and supplies.

Example 8.6   The introduction of ecotones through such project-related activities as landscaping, clear-cutting, and mowing can have a variety of important impacts. For example, mowing highway rights-of-way in a crenulated rather than a linear pattern can significantly increase the amount of shrubby ecotone in a forested area, an ecotone that can provide an enhanced food supply to large browsing mammals such as deer.

Plantations can be used as part of an overall landscaping program to increase the density of selected wildlife populations and even to establish new wildlife populations. Plantations may include indigenous and exogenous vegetative species. In either case, careful consideration must be given to minimize unforeseen and highly undesirable secondary effects. For example, a vegetative species (e.g., blueberry) to be used for extensive landscaping along a highway right-of-way may have been selected primarily for the food value of its berry for selected wildlife populations (e.g., song birds). However, under certain circumstances (e.g., prolonged high summer temperatures, severe drought), the berries of certain plant species may ferment, leading to the intoxication of birds (or other animals), with important consequences to vehicular safety along the highway.

Where ecotones are maintained primarily through significant expenditures of energy (e.g., mowing), a particularly important problem arises when, in periods of severe budgetary constraint, those expenditures cease and the ecotone disappears through succession. In such instances, wildlife populations that were originally enhanced by and have since become dependent on the artificially maintained ecotone can be decimated.

Example 8.7   Impacts on the migratory behavior of animals are typically associated with design features of projects including cuts, fills, and structures (e.g., buildings, fences, roadways) that may act as physical barriers or impediments to diurnal, nocturnal, and diel migrations. However, projects may also introduce other types of barriers to migrations.

For example, area lighting, such as tower lights and spotlights, may interfere with nocturnal migrations whereas noise generated by construction activities (heavy equipment, blasting, drilling) and by operational activities (noises associated with manufacturing processes, traffic) may interfere with diurnal migrations. In some instances, the mere presence of humans precludes the use of otherwise available and desirable habitat by wildlife species that are highly intolerant of humans.

Finally, projects may not only provide barriers or impediments to migrations but also bridges and opportunities. For example, landscaping features, including not only the introduction of selected vegetative species but also the purposely achieved vertical and horizontal interspersion of vegetative types in the project area and its environs, may facilitate or preclude seasonal migrations by providing or negating essential habitat attributes (e.g., distance from open water to overstory, juxtapositioning of dense shrub growth and open grassland).

Example 8.8   Cuts and fills may directly result in changes in existing cool air flow patterns along the ground in the project area and its environs. In some instances, such changes might modify vegetative growth, especially among cold-sensitive plant species that provide food and shelter to wildlife species, including migratory and nonmigratory species. Modifications of the patterns of cool air groundflow might also result in subsequent impacts of more direct social importance, such as the loss of cold-sensitive commercial crops or the channeling of fog into highway rights-of-way.

Example 8.9   Noise associated with the operation of heavy equipment during construction may interfere with the reproductive cycle (e.g., nesting, egg laying, rearing of young) of indigenous and migratory species, with consequent effects on the food webs in which they participate.

Noise associated with the operational phase of certain projects (e.g., highways) may also lead to localized consolidation of soils, due to the persistent sound-induced vibration of the soil column. In some instances, this effect may lead to reduced percolation and subsequent surface erosion.

CHAPTER 9

WETLAND RESOURCES

Although wetlands are among the most biologically diverse and productive resources on the planet, they have historically been considered essentially wasteland fit only to be reclaimed. Wetlands have, therefore, been among the first of a nation's resources to be purposely altered if not obliterated by human exploitation.

The current concern for the enhancement and preservation of wetland resources is based on our increasing understanding of the multitude of functions that wetlands fulfill as highly dynamic ecosystems. Although as many technical definitions and typologies of wetlands exist as there are legal jurisdictions that underlie current efforts to enhance and preserve them, two aspects of wetlands are particularly important for the purposes of impact assessment.

First, wetland ecosystems are physical links between terrestrial and aquatic resources (Figure 9.1). As such, they receive from and contribute to these resources energies and materials that can play important roles in the dynamics and attributes of each type of ecosystem.

Second, because of the dynamic linkage of wetlands to terrestrial and aquatic resources, project-mediated impacts on wetlands may occur not only as direct consequences of a project on a wetland (e.g., filling the wetland) but also as indirect consequences of project impacts on terrestrial and aquatic systems (Figure 9.2).

FUNCTIONS

In 1983, the United States Federal Highway Administration devised a method for the functional assessment of wetlands. This method (Adamus, 1983, *A Method for Wetland Functional Assessment*, U.S. Federal Highway Administration), which has since been adopted by a variety of National

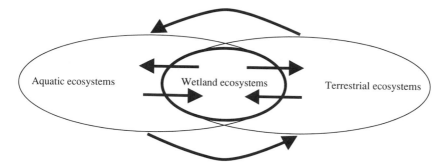

FIGURE 9.1   Wetland ecosystems as dynamic links between aquatic and terrestrial ecosystems. Arrows represent the transfer of materials and energy from one ecosystem to another.

regulatory authorities, includes consideration of the following wetland functions:

- groundwater recharge
- groundwater discharge
- flood storage and desynchronization
- shoreline anchoring and dissipation of erosive forces
- sediment trapping

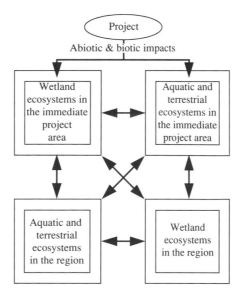

FIGURE 9.2   Project-mediated impacts of wetlands. Impacts include direct impacts on wetlands and other ecosystems in immediate project area, as well as indirect impacts on more distantly located ecosystems.

- nutrient retention and removal
- food web support
- habitat for fisheries
- habitat for wildlife
- active human recreation
- passive human recreation and heritage value

Groundwater recharge is the movement of precipitation or surface water through wetland soils into water table or artesian aquifers (Chapter 7). *Groundwater discharge* is the movement of water from these aquifers into surface water resources, which typically occurs in dry seasons when groundwater may become an important augmentation to seasonably low flowing streams and rivers. *Flood storage* refers to the capacity of a wetland to store peak flows of surface and groundwater (e.g., from storms, snow melt) and, therefore, to delay downstream flooding. *Desynchronization* is the staggered release of peak flows that have been temporarily stored in a number of wetlands in a watershed. Such a staggered or nonsimultaneous release of floodwater from numerous upstream wetlands results in a buffering of downstream floodwater flow, leading to lower but more persistent downstream peak flows.

Shoreline anchoring is the capacity of wetland vegetation to anchor shoreline soils by means of fibrous root systems. Vegetation also acts to dissipate erosive forces (e.g., wind, current, fluctuating water level) that would tend to wash shoreline soils downstream. By serving as a physical barrier to hydraulic flow, wetland vegetation may also serve as the primary means by which suspended particulate matter (sediment) is trapped within the wetland and, thus, is prevented from entering groundwater aquifers or downstream surface waters.

Nutrient retention may occur as a result of the trapping of runoff sediments that contain mineral or organic nutrients, and may also occur as a result of the uptake of dissolved nutrients from runoff water by wetland vegetation. *Nutrient removal* refers to the loss of nitrogenous nutrients to the atmosphere, which may occur during the mineralization of organic matter entrained in runoff into the wetland.

Wetlands may provide significant *food web support* because they directly or indirectly provide food for animals. Such food may derive directly from wetland vegetation (e.g., seed, fruit, detritus) or indirectly, in the biomass of primary consumers (e.g., fish) that serve as a food supply for secondary or higher order consumers (e.g., birds, amphibians, humans). Wetlands may also provide *essential habitat* for the fish and wildlife community in developmental and adult stages of their life cycles. In addition to food web support, other attributes of fish and wildlife habitat mediated by wetlands include temperature, salinity, dissolved oxygen, and the availability of nesting materials.

Active recreational uses of wetlands include those activities that are essentially water dependent, such as boating and swimming. *Passive recreational uses* are those activities that are not directly dependent on access to open water, such as picnicking, aesthetic enjoyment, and nature photography. The heritage value of wetlands refers to their potential importance for the preservation of socially significant features, including archaeological, geological, historical, and biological features.

Although these perceived functions of wetlands are important guides to the assessment of project-mediated impacts on wetlands, the following caveats must be considered.

1. Whether or not a particular wetland fulfills any or all of these functions to a significant degree cannot be answered except on the basis of site-specific data and information. Documentation and procedures for collecting and processing appropriate data and information are available (in both manuscript and computer format) from United States government agencies such as the Federal Highway Administration (FHWA), the Army Corps of Engineers, and the Fish and Wildlife Service.

2. Although these functions were identified and addressed in the development of a specific procedure (originally, the FHWA Wetland Functional Assessment Procedure), such functions represent a broad scientific consensus on possible environmental roles of wetlands.

Specific analytical procedures may, of course, change over time or may be found to be inappropriate with respect to geographical, legal, and site-specific conditions. Therefore, the assessment team should consider it necessary to address these functions and to document the manner in which they are addressed on the basis of current scientific literature. This topic—the use of specific analytical methodologies in impact assessment—is developed more fully in Chapter 10.

Of course, no list of wetland functions, even if it represents a consensus, should be considered definitive for the purposes of impact assessment. For example, none of the functions listed earlier specifically refer an assessment team to the consideration of wetland functions regarding public health (e.g., as habitat for disease vectors) and safety [e.g., drowning and engulfment (by mud) of children].

WETLAND CLASSIFICATION

The functions ascribed to a particular wetland (or to any other natural resource), as well as the values of those functions, essentially derive from an understanding of the ecological dynamics of wetlands. Wetland dynamics vary dramatically with the type of wetland.

Historically, wetland classification in the United States focused on habitat value to waterfowl (Table 9.1). Wetland types such as marshes, swamps, and bogs came into common usage among practicing environmental professionals. More recently, the United States Fish and Wildlife Service created another classification system to encompass a much broader range of wetland values. This new classification system (inclusive of wetlands and open water resources) is a relatively complex, hierarchical scheme based on geographical succession (*provinces*), ecological systems and subsystems, habitat classes and subclasses, and habitat types.

Five major ecological systems are recognized in this hierarchy:

1. **Marine system:** Coastal land and water systems with unobstructed access to the open ocean, including intertidal and subtidal subsystems

2. **Estuarine system:** Coastal land and water systems that are semi-enclosed by land, with open, partially obstructed, or sporadic access to the ocean and a measurable quantity of ocean-derived salt in the water, including intertidal and subtidal subsystems

3. **Riverine system:** Between the channel bank and extending to and including vegetation channelward of the bank, including (a) the high gradient subsystem (area of fast flow, where the stream bed consists of rock, cobbles, or gravel with occasional patches of sand), (b) the low gradient subsystem (area of slow flow, where the stream bed is mainly sand, silt, and clay), and (c) the tidal subsystem (where stream bed gradient is low and velocity fluctuates under tidal influence)

4. **Lacustrine system:** All nontidal habitats situated in depressions that are bound by wave-formed or bedrock shoreline features, or that are at least 20 acres in extent at the deepest portion of the catchment and are devoid of trees, shrubs, or persistent emergent vegetation, including (a) the littoral subsystem (the area from the shoreward boundary to the maximum depth of effective light penetration) and (b) the profundal subsystem (the area in which insufficient light penetration occurs for the growth of rooted or attached macrophytes)

5. **Palustrine system:** All nontidal wetland and aquatic habitats in which water is not restricted to a definable channel, wave-formed or bedrock shoreline features are absent, and persistently nonvegetated and deepest portions of the catchment are less than 20 acres in extent.

Habitat classes, subclasses, and orders are based on dominant vegetation, the form of substrate for nonvegetated area, persistence of leaves (evergreen and deciduous), soil type, substrate texture, substrate origin, and dominant sedimentary-animal communities. An example of the hierarchical

TABLE 9.1   Wetland Types[a,b]

Wetland type	Hydrological features
Inland Fresh Areas	
Seasonally flooded basins or flats	Soil covered with water or water-logged during variable periods; well drained during most of growing season
Fresh meadows	Without standing water during growing season; waterlogged to within a few inches of surface
Shallow fresh marshes	Soil waterlogged during growing season; often covered with 6 inches (plus) of water
Deep fresh marshes	Soil covered with 6 inches to 3 feet of water
Open freshwater	Water less than 10 feet deep
Shrub swamps	Soil waterlogged; often covered with 6 inches (plus) of water
Wooded swamps	Soil waterlogged; often covered with 1 foot of water
Bogs	Soil waterlogged; spongy covering of mosses
Inland Saline Areas	
Saline flats	Soil waterlogged during growing season; at high tide as much as 6 inches of water
Saline marshes	Covered with 6 inches to 3 feet of water at high tide
Open saline water	Shallow portions of open water along fresh tidal rivers and sounds
Coastal Fresh Areas	
Shallow fresh marshes	Flooded after periods of heavy rain; waterlogged within few inches of surface during growing seasons
Deep fresh marshes	Soil waterlogged during growing season; often covered with 2–3 feet of water
Open fresh water	Permanent areas of shallow saline water; depth variable
Coastal Saline Areas	
Salt flats	Soil waterlogged during growing season; sites often covered by high tide
Salt meadows	Soil waterlogged during growing season; rarely covered by tide
Irregularly flooded salt marshes	Covered by wind tides at irregular intervals during growing season
Regularly flooded salt marshes	Covered at average high tide with 6 inches (plus) of water
Sounds and bays	Portions of saltwater sounds and bays shallow enough to be diked and filled
Mangrove swamps	Soil covered at average high tide with 6 inches to 3 feet of water

[a] Reprinted from Shaw and Fredine (1956). "Wetlands of the United States" Fish and Wildlife Service, U.S. Department of the Interior, Circular No. 39.
[b] Although this classification has been superceded by a more recent U.S. Fish and Wildlife Service classification system, these wetland typologies are still used in many places throughout the world.

relationship of habitat order, subclasses, and classes for the palustrine system is shown in Table 9.2.

Depending on the location of the proposed project and relevant legal jurisdictions, the assessment team will find itself using one wetland classification scheme or several. Certainly, for any assessment project conducted in the United States and subject to federal review under the National Environmental Policy Act of 1969 (NEPA), the assessment team is well advised to use the most recent wetland classification scheme available from the Fish and Wildlife Service. For those projects not subject to federal authority that fall under state jurisdiction, the assessment team may need to use state-mandated classification schemes. Similarly, different legal prescriptions and different historical databases will apply to different locations throughout the world.

The essential utility of any wetland classification system to the assessment process is **to identify key factors that influence wetland attributes and dynamics.** Once such factors are known, the assessment team may proceed to relate project design features and activities to those factors. For example, compare the difference in precision between these two questions:

- What are possible ecological impacts of dredging in wetlands near the coast?
- What are possible ecological impacts of dredging in a slightly saline, temporarily flooded, mineral-rich, emergent palustrine wetland in the Acadian Province?

The first question is so general that it gives little if any guidance to the assessment team about the factors, attributes, or dynamics that could be affected by dredging. The second question is much more specific and identifies (in the documentation of the classification scheme used) a range of hydrological, nutrient, and vegetative factors that should be considered in relation to the act of dredging.

ECOSYSTEM DYNAMICS

As in any aquatic or terrestrial ecosystem, the dynamics of wetlands are manifest in the specific and interdependent processes of nutrient cycling and energy utilization. For example, Figure 9.3 is a highly simplified model of an estuarine wetland (Chapter 7) that supports substantial rooted vegetation (vascular plants) as well as free-floating microscopic algae (phytoplankton).

The major energy and material inputs into this wetland ecosystem are (1) sun energy (for photosynthesis), (2) tidal energy (for periodic input of water, flushing out of wastes and distribution of biomass, maintenance of fresh- and salt water regimes, nutrient trapping), and (3) organic matter

TABLE 9.2 Example of a Wetland Classification Hierarchy for the Palustrine Ecological System[a]

	Class	Subclass	Order
Vegetated	Forested wetland	Evergreen	Organic soils / Mineral soils
		Deciduous	Organic soils / Mineral soils
	Shrub wetland	Evergreen	Organic soils / Mineral soils
		Deciduous	Organic soils / Mineral soils
	Emergent wetland		Organic soils / Mineral soils
	Moss–Lichen wetland		Organic soils / Mineral soils
	Floating leaved bed		Organic soils / Mineral soils
	Submergent bed	Vascular plants	Organic soils / Mineral soils
		Algae	Organic soils / Mineral soils
Nonvegetated	Flat		Fine soils / Coarse soils
	Bottom		Fine soils / Coarse soils / Rock / Organic soils

[a] After Cowardin *et al.* (1977). *Classification of Wetlands and Deep-Water Habitats of the United States,* Fish and Wildlife Service, U.S. Department of the Interior.

and nutrients (provided from autochthonous and allochthonous sources). These energies and materials accomplish the fertilization, irrigation, and essential energizing of phytoplanktonic and vascular photosynthesizers and, thus, directly influence primary production. The primary producers (vascular plants and phytoplankton), as well as the organic matter entrained in hydraulic flows into the estuary, contribute to the detritus reservoir within the emergent wetland ecosystem. This detritus reservoir, as in climax communi-

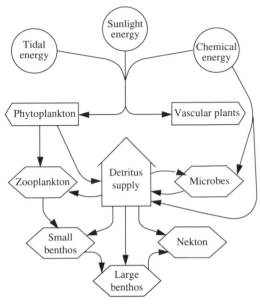

FIGURE 9.3 An example of a simplified systems diagram of an estuarine, emergent wetland ecosystem. Arrows represent the flow of materials and energy.

ties in terrestrial ecosystems (Chapter 8), plays an essential role in the nutrient dynamics of the ecosystem. Detritus is typically the major food supply for a variety of salt-marsh biota, including zooplankton, benthos, and nekton.

Of course, some marine and estuarine ecosystems are not vegetated or do not depend on large reservoirs of detritus (e.g., reefs, rocky shore areas, and flats). The assessment team should be careful that it does not assume that projects can have no ecological impacts merely because of the absence of vegetation in an affected wetland area. Nonvegetated wetlands can be important, for example, for their sequestration of toxic materials in sediments. They can also be important for mineralizing complex organic wastes and exporting released nutrients to other, far-removed vegetated ecosystems. In general, the assessment team should assume that minimal primary productivity in a particular coastal wetland is not necessarily a useful measure of the ecological significance of that wetland. The assessment team should always strive to look at a particular coastal wetland in the context of interconnected aquatic, terrestrial, and other wetland resources.

Inland wetlands (including riverine, lacustrine, and palustrine systems) include a wide range of vegetated and nonvegetated habitats. Some inland wetlands, such as vegetated low-gradient reaches (of riverine systems) and emergent wetlands in littoral areas (of lacustrine systems), also depend on detritus, as do salt water marshes. Other inland wetland habitats, however, are quite different. For example, the riffles and pools of riverine systems and intermittently flooded flats in lacustrine systems are dominated by physical and chemical processes (e.g., mechanical oxygenation of water) rather than biological processes (detritus production) directly linked to vegetative growth.

Key to understanding the basic structure of any wetland ecosystem are the interrelationships that exist among four processes:

- photosynthetic utilization of sunlight by primary producers
- cycling of mineral nutrients
- detritus production and utilization, including export of detritus to downstream communities and sequestration within the wetland
- use of hydrodynamic energy, for example, flow rate and watershed hydrology, lake hydrodynamics, tidal influence, and inundation regime (i.e., periodicity of flooding)

As shown in Figure 9.4, some of these processes may interact to initiate a successional development of the wetland. The particular pattern of succession depicted in this figure begins with the growth of marsh plants,

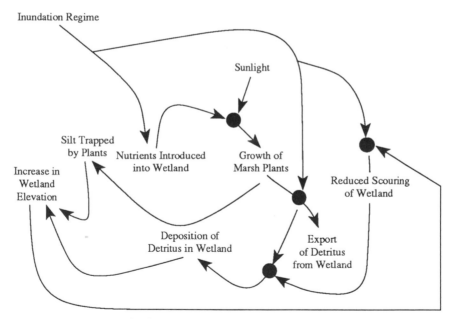

FIGURE 9.4 An example of a negative feedback loop (i.e., increase in wetland elevation) in a marsh that is dependent on periodic floods for inorganic nutrients. Read ● as "a combination that promotes an increase in."

which is basically sustained by nutrients introduced into the wetland by periodic inflows of freshwater (inundation regime). One result of the growth of marsh plants is an increase in silt trapping within the wetland, which tends to increase the wetland elevation. A progressive increase in wetland elevation leads, in turn, to a diminution of overland inflow into the wetland, a consequent reduction in the hydraulic scouring of wetland detritus, and, thus, an increase in the deposition of detritus within the wetland at the expense of external export to downstream receiving waters. In this case, vegetation-mediated silt trapping and autochthonous deposition of detritus constitute negative feedback loops on the wetland's inundation regime, resulting in a wetland ecosystem that is progressively less dependent on an externally driven hydrologic regime and allochthonous nutrient supply.

As in aquatic and terrestrial ecosystems, periodic physical perturbations that result from forces of conditions not subject to the control of ecosystem dynamics can play important roles in the stability of wetland ecosystems. An example in aquatic systems is the season-mediated breakdown of thermal stratification in a lake, resulting in lake overturn and the periodic fertilization of surface waters (Chapter 7). An example in terrestrial systems is the periodic occurrence of fires, which typically returns a successional sere to an earlier successional stage (Chapter 8). Ecosystems that tend to manifest key attributes and dynamics as a result of such periodic physical perturbations are often referred to as *pulsed ecosystems*. Many wetlands are essentially pulsed ecosystems whose salient characteristics are directly dependent on the periodicity of hydrologic flow. A good example is, of course, the saltwater marsh, which is dependent on tidally pulsed irrigation and nutrient recycling.

Other types of pulsed wetlands are less obvious. For example, shallow-water wetlands marginal to lentic and lotic systems may quickly become anaerobic and acidic, resulting in a diminished decomposition of vegetative debris washed into them or produced by wetland vegetation. Thus, the bottom deposits of such wetlands tend to become rich storehouses of organic material that progressively fills the wetland, displacing standing water and eventually leading to terrestrial successional patterns in which deeply rooted vegetation, acting as a nutrient pump, transforms deeply sequestered nutrients into surface biomass.

Periodic flooding of these areas, especially by floods with high scouring potential, can transport the organic nutrients trapped in soils and inundated vegetation to downstream sources. Such pulsed fertilizations of downstream receiving waters, occurring seasonally (e.g., during freshet flow in temperate areas, during monsoon in tropical areas) or over extended periods of years, may play a key role in the maintenance of downstream communities. The hydrologic pulses themselves, of course, also return the successionally developing wetlands to an earlier successional stage.

When hydrologic pulses become important to the long-term stability of wetland ecosystems, human-mediated influence on those hydrologic pulses (e.g., through flood control measures, watershed development) can clearly result in significant impacts on the long-term stability of not only the wetlands but also the downstream ecosystems to which they are connected.

Note that interference with both high frequency (e.g., seasonal) and low frequency (e.g., decade) hydrologic pulses in a wetland may lead to long-term ecological impacts within and outside that wetland. For example, seasonal dry periods (a relatively high frequency event) that reduce a wetland's water level may also concentrate fish populations on which a migratory bird species depends. A human-mediated increase in water level (e.g., due to watershed development) during the migratory season, resulting in a dispersion of fish populations, may effectively remove this food supply, with potentially long-term impacts on the avian population and on distantly located ecosystems of which that population is an important component.

A special issue that has come into prominence in the past decade with respect to wetland functions and dynamics is that of *replacement wetlands*. Replacement wetlands are wetlands that may be purposely constructed as replacements for wetlands that are filled or otherwise affected by project development. These replacements may also be wetlands that already exist outside the immediate project area but are legally procured by the developer and then ceded to an authority for the express purpose of preservation. Again, the attempt is to provide a replacement for wetlands "lost" to project development. In either case, regulatory authorities typically impose a requirement on the developer to provide an amount of replacement wetland that is greater than the amount of "lost" wetland. For example, in the United States, the proportion generally ranges from 3 to 7 acres of replacement wetland for each acre of "lost" wetland.

Although replacement wetlands are generally defended as a practical means of mitigating project impacts, clearly the vast majority of wetland "replacements" accomplishes political goals rather than ecological goals. For example, wetland functions and dynamics are typically not solely dependent on a single variable (such as acres of surface area), but on a constellation of physical, chemical, and biological variables. The functions and dynamics of a wetland are, however, quite typically and directly influenced by site-specific conditions that are not likely to be precisely or even significantly duplicated in different locations.

This argument does not suggest that achieving an essential equivalence between the functions and dynamics of one wetland and another is not possible, but that, in the absence of meticulous study, it is highly improbable. Just as wetlands are not wastelands fit only to be filled, they are not simply "wet holes in the ground" that can be moved about from one place to another with little consequence. Wetlands should be viewed as spatially and temporally unique interfaces of dynamic terrestrial and aquatic resources that not only reflect specific attributes of those resources, but also contribute to and influence those attributes.

In light of these considerations, clearly the assessment team must take a comprehensive systems overview of coastal and inland wetlands. Such an overview, which is in direct opposition to the historical perspective of wetlands as wastelands, should encompass the following considerations:

- All wetlands, regardless of location or type, perform some type(s) of work within the physical environment.
- The work performed by a particular wetland may directly or indirectly influence local and distant ecosystems, including other wetlands as well as aquatic and terrestrial ecosystems.
- The work performed by a particular wetland and, thus, its dynamic interrelationships with other ecosystems can be at least qualitatively inferred from (1) the specific types of energy and material inputs into the wetland, (2) the water–soil–biological attributes and dynamics of the wetland, and (3) the specific types of energy and material outputs from the wetland.

EXAMPLES OF DIRECT AND INDIRECT IMPACTS

Because they are interfaces of land and water, many examples previously discussed with respect to terrestrial and aquatic resources also apply to wetlands. The following examples have been selected to highlight some of the unique features of wetlands and, in some instances, to emphasize the dynamic relationships among wetland, aquatic, and terrestrial ecosystems.

Example 9.1 With the exception of *perched wetlands* (i.e., where the topological gradient descends along the entire periphery of the wetland), wetlands act as hydrologic sinks that receive terrestrial runoff. Included in this runoff may be not only nutrients, organic particles, and sediment but also toxic chemicals, which may be dissolved in runoff water or adsorbed to entrained particulates.

Toxic chemicals that may accumulate in wetland bottom muds and peat may derive directly from a variety of projects and human activities, including the construction of highways, with consequent runoff of metallic dusts and hydrocarbon emissions that may fall out within the watershed, as well as toxics released as a result of vehicular accidents; development of agricultural land, with consequent runoff of pesticides; management practices associated with pipelines and transmission lines, such as the use of herbicides within rights-of-way and on-site storage of toxic materials and supplies; and the placement of chemically contaminated fill during project construction. Other more indirect means of toxics introduction into wetlands include topological changes as a result of project development, which may result in altered surficial or groundwater flows through previously sequestered soil contaminants into wetland catchments, and project-related improvement in access to wilderness areas, with consequent increase in hunting, which can result in the introduction of lead shot into wetland muds.

When affected wetlands serve as important recharge areas for groundwater aquifers used as potable supplies, toxic chemicals derived from those wetlands may adversely affect human health. This event would be of special concern with respect to heavy metals that might become mobilized (solubilized) by the acidity (e.g., pH <5.0) of certain wetland waters. Within the wetlands themselves, as well as in downstream waters that receive wetland discharge, toxic chemicals may become biomagnified (see Ex. 7.2, 8.5), with consequent impacts on wildlife populations and the grazing and detritus food webs associated with those populations, including food webs in which humans participate as higher order consumers

Example 9.2 Although wetlands may be highly productive with respect to wildlife and commercially important species, they also may provide excellent habitat for a variety of human disease vectors such as rodents and mosquitos. The purposeful construction of a wetland (e.g., as "replacement wetland") as a means of mitigating project impacts such as filling existing wetlands along a proposed highway right-of-way may result in an enhanced health risk to persons living in the immediate area of the new wetland. When the replacement wetland is located to serve as an "attractive nuisance" to children, the so-called "mitigation" effort might result in an increased safety risk for children because of the potential for drowning and engulfment by deep muds.

Example 9.3 The purposeful introduction of new wetlands as well as the maintenance of existing wetlands (e.g., legal acquisition for purposes of preservation) that serve as prime habitat for waterfowl, **but are located in the immediate vicinity of a lacustrine resource,** is likely to result in (or prolong) the avian fecal contamination of that resource. The direct impact on the lake or pond may be to enhance its eutrophication (see Ex. 7.4), with possible consequent impacts on primary contact recreation, the economic value of shoreline property, and water potability.

Example 9.4 Upstream channelization of a river or stream will typically result in increased flow velocity in downstream reaches. Riverine wetlands in such downstream reaches will subsequently become subject to increased scouring, with potential loss of benthic habitat that may provide substrate and detrital nutrient to detritus-based food webs. Changes in the hydraulic gradient in such wetlands, as well as loss of benthic habitat and detritus, could have long-term adverse impacts on riverine fisheries that might use the wetlands as nurseries or as resting or food supply areas. A similar causal chain of events can result from other upstream activities that may increase flow rate, including land clearing and the placement of impervious surfaces (e.g., bitumen parking lots, road surfaces, roofs of buildings).

A decrease in flow velocity in downstream wetlands may result from such project-related activities as the construction of dams (e.g., for flood control), interbasin transfers of water, diversion of upstream flow for irrigation or for human consumption, and upstream excavation and landscaping that diverts surficial or groundwater flow to other receiving waters. Decreased flow velocity in downstream wetlands may result in significantly altered wetland habitat (i.e., depth of water, temperature, concentration of oxygen, sunlight penetration) and, by reducing the rate of scouring, promote a new pattern of wetland succession.

Example 9.5 The filling of a wetland that serves as an important recharge area to groundwater aquifers will directly result, of course, in the loss of that recharge capacity and, indirectly, in the drawdown of affected aquifers. Alternatively, the filling of a wetland that, because of its underlying geology, does not serve as a recharge area but as a means of water loss through evaporation will directly result in the local conservation of groundwater and, indirectly, in the enhancement of the on-going uses of that groundwater.

Regardless of the hydrologic impacts of filling a wetland, filling will result in a loss in the wetland's capacity to perform a variety of other potential functions, including flood storage and desynchronization, food web support, wildlife and fisheries habitat, and support of active and passive human recreation.

The filling of a wetland that intercepts sheet flow into a lacustrine, riverine, or estuarine system may, depending on the physical and chemical constituents of that sheet flow, result in the increase in turbidity and concentrations of nutrients and/or toxic chemicals in that lacustrine, riverine, or estuarine system. Similarly, the purposeful construction of wetlands to intercept sheet flow can result in a decrease in turbidity and concentrations of nutrients and/or toxic chemicals in subsequent receiving systems.

Example 9.6 Shallow warm-water palustrine wetlands that are overgrown with floating vegetation are typically oxygen deficient and sometimes highly acidic. Depending on the relative volumes or flow rates in these wetlands and the interconnected lacustrine or riverine systems, such wetlands may measurably influence the pH and concentration of oxygen in these latter systems, especially during seasonably dry periods. In such instances, the filling of these adjacent wetlands may improve the oxygen budget of the lacustrine or riverine receiving waters, as well as reduce their acidity.

CHAPTER 10

SPECIAL ISSUES: PHYSICAL ENVIRONMENT

Throughout the development of the impact assessment process as a tool in decision making, many techniques have been developed to assist in the identification and the valuation of potential impacts. Some of these, such as overlay, checklist, matrix, and network techniques; habitat evaluation procedures; and wetland functional assessment techniques (Chapters 3,9) have been used extensively and are considered standard procedures within the assessment community. Other techniques have not been used as extensively but are nonetheless potentially powerful tools that might be considered on a project-by-project basis. These alternative tools include bioassays, modeling techniques, and hazard analysis.

BIOASSAYS

Bioassays are experimental procedures designed to show the biological effects of changes in the physical or chemical conditions to which a test organism or population is subjected. These tests can be used for a variety of purposes coincident with assessment needs, including the determination of:

- the suitability of environmental conditions for biological growth and well-being,
- favorable and unfavorable concentrations or values of specific environmental factors,
- the toxicity of chemicals, and
- the nutrient value of chemicals.

Standard bioassay procedures relevant to the needs of impact assessment are available for a wide range of aquatic species, including phytoplankton, zooplankton, coral, annelids, crustaceans, aquatic insects, mollusks, and fish. Relevant bioassays for terrestrial species are less numerous but include some that might be pertinent to the assessment of certain types of impacts on herbaceous vegetation, including the effects of sodium and mercury vapor lighting, salt runoff from highways, soil moisture and temperature, and air emissions.

Although bioassays are experimental procedures, they are typically designed to avoid the need for expensive laboratory apparatus and can be performed by personnel who are not professional scientists. Many of these procedures can be carried out in the field as easily as in a laboratory. An example of how a bioassay may prove useful for purposes of impact assessment follows.

Consider that the assessment team has determined that the projected springtime clear-cutting in a project area upstream from a lake will result in periodic sheet erosion and that soil particles entrained in the runoff will be carried into the downstream lake. The team is concerned that these soil particles might contain sufficient nutrients to enhance the growth of algal mats significantly, which certainly would adversely affect the popular use of the lake for swimming.

Of course, analyzing the soils in the project area to determine their nutrient content and, by performing various calculations regarding the particulate loading of the lake, generating data that may or may not substantiate the concern would be possible. However, such an approach is riddled with many difficulties, not the least of which might be the limited knowledge of the specific nutrient requirements of the algae. A more direct approach would be conducting a bioassay, which might be described simply as a comparison of (1) algal growth in lake water that has no addition of an extract derived from entrained soil particles (called the "control") with (2) algal growth in lake water that does have an addition of extract derived from entrained soil particles (called the "test"). As actually carried out in a laboratory, such a bioassay would include a number of tests, each having a progressively greater concentration of soil-derived extract added to lake water.

Figure 10.1 depicts some results that might be obtained from such a bioassay conducted in a laboratory. In this case, progressively greater concentrations of soil-derived extract clearly do not enhance algal growth relative to the untreated lake water. Thus, the experiment provides a reasonable (and, if done correctly, replicable) basis for deciding that the proposed impact (i.e., enhanced algal growth) is highly unlikely. In fact, because the data provided in Figure 10.1 show that algal growth is progressively decreased with increasing concentrations of soil-derived extract, the extract is likely to contain chemicals that are toxic to the tested algae. Such a result would, of course, raise additional concerns and prompt additional assessment effort.

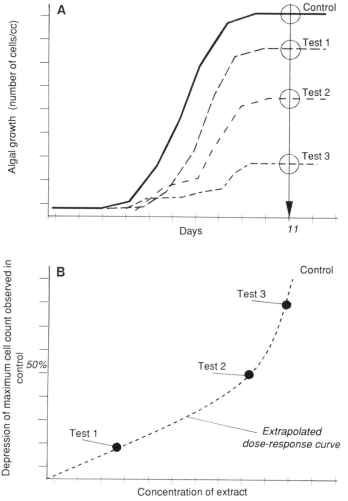

FIGURE 10.1 Hypothetical data derived from an algal assay to determine the toxicity of soil particles released into an aquatic ecosystem by soil erosion at project site. (A) Algal cell counts compared after 11 days, when all cultures have reached maximum concentration. Cell count in lake water not having any addition of extract (Control) is higher than cell counts in lake water to which progressively greater concentrations of extract were added (Tests 1, 2, and 3). (B) Increasing depression of cell counts with increasing concentration of extract shows a dose–response relationship. Such a dose–response relationship is a strong indicator of the possible toxicity of the extract.

As this example demonstrates, bioassays are useful not only for answering certain specific questions (e.g., will soil particles enhance algal growth in the lake?) but also for identifying potential impacts (e.g., what do the demonstrated toxic effects on algae imply regarding ecosystem dynamics and public health?). Also important to the assessment effort is the fact that bioassays can be conducted at very low cost. For example, the bioassay just described might easily be conducted (from the start of the experiment through the analysis of data) at the expense of 2–3 person-days over a period of 10 days.

Of course, the expense of a bioassay (in terms of money and time) will vary with the type of assay conducted. Assays involving a single species and a single environmental parameter are typically much less expensive and time consuming than studies that, in an attempt to duplicate actual ecological dynamics, involve a variety of interacting biotic species and a range of physical and chemical conditions. These more ecologically oriented studies, often subsumed under the heading of *environmental effects testing,* may generally be described in terms of the volumes of growth and test media required, as well as the time to conduct the study. The following environmental effects testing techniques are generally useful, but a specific type of study may vary greatly from one application to another.

1. **Microcosm (laboratory) studies:** Involve small volumes of growth and test media (e.g., < 1 liter) and require relatively short periods of time (e.g., <2 weeks); physical and chemical parameters are rigidly controlled
2. **Mesocosm studies:** Involve on the order of 1–10 liters of growth and test media and require up to several weeks or more; may be conducted under laboratory or field conditions, with consequent variability in the control of physical and chemical parameters
3. **Macrocosm (large-scale field) studies:** Involve typically hundreds of liters of growth and test media and require up to several months; physical and chemical parameters typically subject to ambient seasonal and weather conditions

Specific protocols for standard bioassays and environmental effects testing are available from a variety of sources, including the United States Environmental Protection Agency, the Organization for Economic Cooperation and Development (OECD), the American Society for Testing and Materials (ASTM), and the American Public Health Association (APHA).

MODELING TECHNIQUES

Physical and mathematical models have been used in an attempt to quantify a variety of potential environmental impacts.

Historically, physical models have been used in assessing impacts of relatively large-scale projects, such as the mixing dynamics of interbasin transfers of water and the hydrodynamics of port development. Because of the high cost typically associated with the design, construction, and operation of physical models, because of technical difficulties involved in extrapolating data derived from relatively small physical models to actual large-scale design features and dynamics, and because of the explosive growth of computer technology, physical models have been essentially replaced by computerized mathematical models as cost-effective flexible tools for the purposes of impact assessment. Currently, hundreds of computerized models are available in the technical literature, many of which may have relevance to a particular type of impact, including the estimation of:

- the environmental pathways and fate (Figure 10.2) of chemicals, including organic and inorganic chemicals,
- the human exposure potential of chemicals,
- the accumulation of chemicals in biota, and
- the persistence and degradation of toxic chemicals in soil and water.

Given the large number of available models, the diversity of their possible applications, and the ready availability of appropriate computer

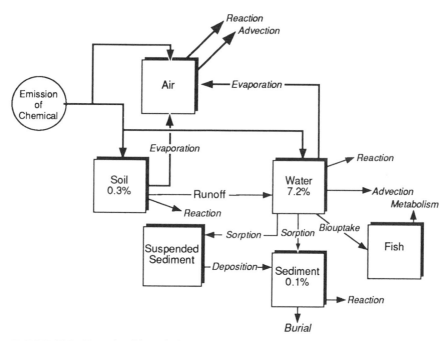

FIGURE 10.2 Fate of trichloroethylene in a six-compartment system. Adapted by permission of the Canadian Water Resources Association from D. Mackay (1987). The holistic assessment of toxic organic chemicals in Canadian waters. *Can. Water Res. J.* **12**: 14–20. Copyright 1987.

hardware and software, the following caveats regarding the use of any mathematical model for the purposes of impact assessment are important.

1. Any mathematical model of environmental components and dynamics is (and must be) a simplification of actual environmental components and dynamics, because of the various assumptions that must be made to construct the model. No mathematical model should be used for purposes of impact assessment until the assessment team has thoroughly examined these assumptions and has determined that they are consistent with assessment objectives and site-specific conditions.

2. The use of any mathematical model is essentially the use of an analytical technique that must, as must any technical protocol, be validated in terms of a **verification procedure.** Such a procedure must address a range of issues that ultimately determines the usefulness of the model, including such issues as the statistical limits of precision, the sensitivity of model output to model input, replicability of outputs, and field or laboratory validation of outputs. Some of the issues to be addressed in a verification procedure are included in Figure 10.3. As shown in this figure, verification is typically a multiphased effort requiring progressively detailed comparison of computed predictions with actual environmental conditions.

3. The assessment team must consider that computerized mathematical models available for the purposes of impact assessment (a) are quantitative and (b) typically focus on physical, chemical, and biological dynamics. As discussed in Part I, impact assessment is inclusive of nonquantifiable as well as quantifiable attributes of the environment, and must also consider social components and dynamics. The assessment team should, therefore, consider whether the use of any mathematical model, in terms of expenditures of time and money, will promote or mitigate against a balanced assessment of the total environment.

Certainly computerized models that are useful for estimating human exposure to potentially toxic chemicals associated with a proposed project may provide important insight to potential impacts on human health and, where appropriate, should be seriously considered. However, the use of models that require very substantial effort but nonetheless narrowly focus on the physical environment may substantially subtract from the assessment team's capacity to achieve a balanced consideration of the physical and social environments.

4. In many instances of impact assessment, the data requirements for (or the assumptions of) a particular model cannot be met or can be met only through significant expenditures of time or money. In such instances, the assessment team might reasonably consider a **worst-case approach.**

For example, consider that an assessment team is considering the possible effects of highway runoff on a receiver stream. Concern has been

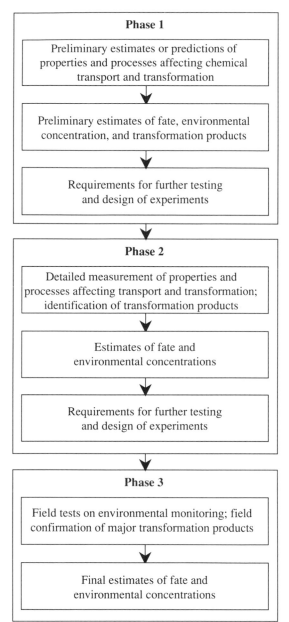

FIGURE 10.3 Sequence of environmental fate/chemistry testing. After Pearson and Glennon (1979). In *Analyzing the Hazard Evaluation Process*, (Dickson, Maki, and Cairns, Jr., eds.), American Fisheries Society.

expressed that salt applied to the highway during winter will adversely affect the fisheries in the receiver stream. Under such circumstances, the assessment team may attempt to model the flow of salt-containing runoff through watershed soils and ultimately into the receiver stream. This model would require careful consideration of a variety of factors, including seasonal precipitation and flow rates of streams, the ion-exchange capacity of watershed soils, and variable salt-application rates.

Another approach is to assume that all the salt applied to the highway during the winter will enter the stream. Alternative worst-case calculations could also be performed by making different assumptions about the timing of salt release into the receiving stream, for example, all at once or over some interval of days or weeks. Although such worst-case calculations are certainly unrealistic, they can be useful short-cuts to determining whether or not any possibility exists for the imagined impact to occur. In the preceding example, if the worst-case calculation of salt-loading the stream (i.e., assuming that all salt goes into the stream at the same time) demonstrates that the resultant chloride concentration in the stream would be well within the tolerance limits of the fisheries of concern, no further assessment would be required.

HAZARD ANALYSIS

In recent years, several techniques have been devised that are often collectively referred to as *hazard analysis techniques*. These techniques have been designed, primarily in response to the need to control the release of hazardous chemicals from manufacturing facilities to the environment (e.g., see *Process Safety Management Regulations*, U.S. 29 CFR (1910.109), to answer one or more of the following questions:

- What can go wrong?
- What are all the causes?
- How bad can it be?
- What should be done about it?

Although these techniques are typically used to assess plant design features and operations, many may be adapted for purposes of environmental impact assessment.

Relative Ranking

Relative ranking techniques are used to compare the attributes of several manufacturing processes or activities (1) to determine whether they pose enough of a hazard to warrant additional study and (2) to evaluate alternative means for reducing or managing risk. Comparisons are made on the basis of numerical values that signify the analyst's judgment of the significance of hazard.

With respect to impact assessment, this approach is particularly relevant to the scoping phase of impact assessment in which priorities may be assigned to potential impacts on the basis of the type of impact—public health hazard, loss of local economic base, change in local land use, and so on. Priorities may be assigned not only on the basis of type of impact, but also with regard to phase of project development (i.e., pre-construction, construction, and operational phases) or even relative duration of impact (i.e., days, months, years, decades).

Note that the application of relative ranking techniques to impact assessment does not result in technical assessment of the possibility of impacts or their causal chain. Instead, the primary use of this technique is in assigning priorities among diverse assessment efforts, an essential prerequisite to using impact assessment as an iterative process of project decision making.

Preliminary Hazard Analysis

A *preliminary hazard analysis* (PHA) is a technique that focuses on the hazardous materials and major processing areas of a plant to identify hazards and potential accident situations. This technique requires consideration of plant equipment, the interface among plant components, the operating environment, specific plant operations, and a description of the physical layout of the plant. The purpose of this technique is to assign a criticality ranking to each hazardous situation that may be envisioned, even in the absence of specific information about the actual plant design features or operating procedures.

With respect to environmental impact assessment, this technique is particularly relevant to the early phases of assessment, when broadly defined causal chains (e.g., excavation, which leads to suspended particles, which leads to possible chemical contaminants, and so on) can be identified and can guide subsequent analytical efforts.

What-If Analysis

In *what-if analysis* (WIA), experienced personnel formulate a series of questions that must be evaluated with respect to potential hazard. Typical questions include "What if pump 23-b shuts off?" and "What if the operator forgets to empty the overflow tank at the end of the week?".

With respect to impact assessment, WIA is a key tool for identifying potential impacts by means of defining possible causal chains. For example, *what if* air pollutants generated from the proposed project settle out in the watershed of a surface water supply used as a potable water supply?

In using this approach, the assessment team should explicitly consider two additional questions that are implicit in the question "What if?": "So what?" and "Says who?". For example, *what if* particulates released during excavation enter a stream? The direct impact would be an increase in the turbidity of the stream. *So what?* To answer this question, the assessment team would have to identify the consequences (i.e., secondary and higher order impacts) of enhanced turbidity. For instance, the enhanced turbidity may lead to occlusion of sunlight. Again, *so what?* What are the possible consequences of occluding sunlight with respect to plants, water temperature, and other features of the project area?

The result of a persistent posing of the question "So what?" is a chain or network of interrelated impacts. As the chain or network expands, physical impacts of project development are highly likely to be seen to have ramifications for social components and processes. Thus, turbidity may influence the well-being of aquatic plants and fish, but turbidity can also influence the recreation and aesthetic use of streams and rivers.

Each answer to the question "So what?" should be immediately addressed by the question "Says who?". The assessment team should attempt to verify each component of the causal chain. Such verification may be on the basis of personal experience, professional opinion, or documented authority. In many instances, answering the question "Says who?" adequately will prove difficult. Such instances are nonetheless important, because they identify elements of a proposed causal chain that must be investigated more carefully.

Hazard and Operability Study

A *hazard and operability study* (HAZOP) typically requires detailed information on the design and operation of the facility of concern. Using this technique, an interdisciplinary team uses a standard set of *guide words* that, when combined with specific *process parameters,* lead to *resultant deviations* to be studied in greater detail with respect to health and safety goals. For example, the guide word "less" might be combined with the process parameter "pressure" to produce the resultant deviation "low pressure." The interdisciplinary team then focuses on (1) the possible causes of "low pressure" (e.g., in a processing tank) and (2) the possible consequences of "low pressure" (e.g., slowing in rate of chemical reaction).

With respect to impact assessment, this technique is particularly useful for establishing possible causal chains in environmental networks composed of physical, chemical, biological, and social components and dynamics and for assessing the probability of impacts. Appropriate guide words for impact assessment might include "increase," "decrease," "turn on," "turn off," "remove," "add to," "magnify," and "replace"—words that denote relatively direct consequences of project-related activities, as in clear-cutting "turning on" light at the previously shaded forest floor. Appropriate process parameters might include features that are also appropriate for manufacturing processes (e.g., temperature, pH, flow) as well as features more typically associated with dynamic environmental systems (e.g., concentration of oxygen, rate of photosynthesis, supply of detritus, biomass).

Failure Modes and Effects Analysis

Closely related to the WIA, the *failure modes and effects analysis* (FMEA) considers the various failure modes of specific equipment and the effects of such failures on plant operation. For example, what if a specific control valve (1) fails in the open position, (2) fails in the closed position, or (3) leaks in the open position.

The primary relevance of this type of analysis to environmental impact assessment is in the design and evaluation of mitigation techniques, which include managerial and engineering means for minimizing adverse impacts and maximizing beneficial impacts (Chapter 19). For example, assessors may decide that fencing can be used to keep large mammals out of a highway right-of-way. In this case, fencing is a means of mitigating roadkill. The objective of a failure modes and effects analysis would be to identify possible failures in the proposed fencing that may be due to its location, height, and maintenance, as well as other causes.

Fault Tree Analysis

In *fault tree analysis* (FTA), a specific accident or plant failure is defined (e.g., release of a toxic gas) and all design, procedural, and human errors leading to that event (called the "top event") are graphically modeled in a "fault tree." The fault tree allows the analyst to define and rank particular groupings of external factors, equipment failures, and human errors (called minimal cut sets) that are sufficient to lead to the top event.

This technique is particularly useful in early phases of impact assessment when the objective is to relate project activities to types of impacts. For example, biomagnification of toxic chemicals is increasingly recognized as an impact that should be considered in the assessment of many (if not all) projects. Using biomagnification as a top event, the assessment team may begin by constructing logically connected events and conditions (i.e., causal chains) that might lead to biomagnification. One branch of these causal chains may be based on project-generated particulates that may contain toxic chemicals, for example, fugitive dust (during landclearing), suspended particles (during excavation), and air emissions. Another branch may be related to on-site storage of materials, use of pesticides, and accidental spills or releases to the environment. Using fault tree analysis in this manner requires constructing the tree out of failures and errors such as "the failure to test fill material for toxic chemicals" or "the failure to control fugitive dust during construction."

Depending on the manner in which the fault tree is constructed, FTA may also be used as a tool for evaluating impacts, especially with respect to probability of occurrence, and for evaluating proposed mitigation measures.

Event Tree Analysis

Whereas an FTA focuses on failures in equipment or personnel that lead to the top event, *event tree analysis* (ETA) includes successes and failures that result in an accident or in the system returning to a safe state. This type of analysis is typically used to analyze complex processes that have several layers of safety systems or emergency procedures.

As can FTA, ETA may be used in early phases of impact assessment to project activities to selected impacts, to compare relative probabilities of occurrence among impacts, and to evaluate proposed mitigation measures.

ADEQUACY OF THE ASSESSMENT EFFORT: GENERAL GUIDELINES

The following general guidelines are offered as a means of testing the adequacy of the assessment effort as it pertains to the physical environment. These guidelines do not pertain to any specific technique or methodology of assessment but to the overall organization, management, and conduct of the assessment process.

1. All phases of the proposed project should be considered, including early systems planning, design, location, acquisition, construction, and operation/maintenance phases.

2. All project activities in each phase of project development (e.g., blasting, clear-cutting, mowing, relocation of residents) should be identified and evaluated for potential impacts on the physical environment.

3. The timing and duration of each project activity should be related to other important events and activities in the general project area and its environs, including seasonal changes in meteorology and hydrology, animal migrations, and patterns of recreational and other uses of natural resources.

4. Individual abiotic and biotic components of the physical environment in the project area must be identified. Dynamic interrelationships between individual components must be defined on the basis of scientific and technical principles and concepts.

5. Key factors (including physical, chemical, biological, and social factors) that can influence dynamic interrelationships must be identified and specifically related to the proposed project and associated activities.

6. The past history of the physical environment in the project area and its environs must be examined and short- and long-term projections be made of the likely future development of the physical environment, with and without the proposed project.

7. All potential changes in the physical environment as a result of the proposed project should be clearly distinguished from changes that would occur in the absence of the proposed project.

8. Mitigation measures should be identified as early in project development as possible and should be evaluated for their feasibility and efficacy.

9. Individual project impacts on the physical environment in the project area and its environs should be examined for consequent effects on distantly located ecosystems.

10. All governmental agencies, universities, and other organizations having regulatory, scientific, and/or technical interest in the physical environment of the general project area should be identified and utilized as important informational sources.

11. Individual local residents having special expertise and experience with respect to the local physical environment should be identified and utilized for data gathering and/or evaluation efforts.

12. Literature surveys should be conducted for the purpose of identifying issues and findings relevant to the physical environment of the proposed project area.

13. When field, laboratory, and/or computerized modeling studies are undertaken as part of the assessment process, only standard, professionally recognized methods and procedures should be utilized.

14. Conclusions should be based on previously tested principles and hypotheses, not on developing state-of-the-art research.

15. Quantitative techniques should be employed when feasible and relevant (in terms of their assumptions). Subjective interpretations and valuations should be clearly identified as such.

PART III

THE SOCIAL
ENVIRONMENT

OVERVIEW OF COMPONENTS
AND DYNAMICS

Although much disagreement exists over how individual social systems specifically function, and over the relative importance of personal, interpersonal, and institutional attributes, these key components of social systems are generally agreed on:

- interacting "actors," that is, individuals who interact with one another **within a system of mutual expectations of one another's behavior,**
- a physical or environmental situation or context in which actors interact with one another,
- motivations of individual actors, and
- a cultural context in which expectations, motivations, and the physical environment are defined by means of **socially shared symbols.**

The word "culture" has a meaning to social scientists that is quite different from its usual lay meaning as "pertaining to the arts." To the social anthropologist, the culture of a people has historically meant a set of behavioral patterns such as "customs," "traditions," or "habits." More recently, the word "culture" has increasingly come to refer to specific rules for the governing of behavior.

The concept of culture as a set of rules that govern behavior has been influenced largely by the development of computer technology, which has permitted comparisons between the computer (as an active information processor) and the human brain. No computer, regardless of the sophistication of its design, can process any information without a program, that is, a set of directions that is fed into the computer and tells it what to do and when and how to do it. Metaphorically, the human brain may be viewed as a highly sophisticated information-processing machine that also requires programming in order to function. The set of programs given to the human computer is culture; the process by which the human computer is given its programs may be referred to as the *enculturation* (in anthropology) or the *socialization* (in sociology) *process.*

The underlying assumption in the concept of culture as a set of programs is that the human being is born essentially in a state of ignorance about what it should do, how to do it, or why. The enculturation or socialization process is the process by which the biological being becomes a social being, knowing who it is, who others are, and what the relationship between self, others, and the rest of the world is and ought to be. Although interest in the role that genetic inheritance may play is increasing, cultural inheritance is still considered to be the vastly more important determinant of human behavioral patterns.

Much of the legislation and regulatory literature pertaining to impact assessment uses the term "culture" in a very different sense than is typically understood by professional social scientists. In this literature, cultural resources are generally considered to be sites, structures, objects, and districts significant in history, architecture, archaeology, or culture. In other words, cultural resources are physical entities.

Although the assessment team must fit its assessment to relevant regulatory constraints, it must also recognize the disciplinary meaning of the concept of culture in diverse social science disciplines and contexts; whether the concept of culture is utilized (as it has been) to refer to particular customs and rituals, to particular personality types, to ethnic enclaves, or to street gangs, or (more generally) to refer to rules governing human behavior, the concept of culture is a fundamental concept of social science. Insofar as the assessment team must deal with professionals and professional literature in the social sciences, the assessment team should not restrict its consideration of cultural impacts to deliberations of project impacts on physical entities used by humans. **Impacts on culture are impacts on the very mechanisms by which humans learn to be the people they are.**

To the individual who is not a social scientist, implying that highways, power plants, dams, and all the other types of projects that typically require impact assessment can actually impact on mechanisms by which people learn to be the people they are might seem presumptuous. Our ignorance of the mechanisms involved makes such social impacts appear far-fetched. Just as the assessment team can reasonably assume that we do not understand everything about the physical environment, so it must also assume that we do not understand everything about our social environment. After all, the fact that we are all physical, chemical, and biological entities does not insure our understanding of the physics, chemistry, and biology of life. Neither does the fact that we all live in social systems insure our understanding of those systems.

CULTURE AS RULES GOVERNING BEHAVIOR

Essentially, culture is what we learn as opposed to what we inherit genetically. What we learn from our experience with the social world around us and from our experience with the physical world around us—especially as it is explained to us by others—constitutes culture, as does what we learn by our perceptions of the usefulness of socially perpetuated knowledge in meeting the real and perceived necessities of our daily lives. Culture is *experiential* because it is derived from and mediated by our and others' experience. Culture is *transmittable* because the experiential knowledge of one group or generation can be passed on to another group or generation. Culture is also *directive* because the transmitted experiential knowledge of one generation influences the experiences of a succeeding generation. Finally, culture is *variable* because the transmission of cultural knowledge from one generation to another is imprecise and because the environment that supports culture may vary.

We can envision the generalized concept of culture by imagining culture as a set of "social dies" that are composed of previously tested behavioral patterns and that continually give "social shape" to the biological and genetic potential of each new individual. Essential components of these dies include human experience of *self, others,* and the *physical environment.* Understanding that psychological, social, and environmental factors that influence cultural rules can change is important. Thus, cultures can be dynamic changing sets of rules as well as static sets of rules. Understanding that "a culture" is different from "a society" is also important. The difference between these two concepts has been likened to the difference between content and form. The *cultural content* of social life includes those rules governing *behavioral patterns* and *ideological* (thinking and believing) *patterns.* The *social form* is the manner in which these rules are actually manifest in the organization of social life. Thus, a society is the means by which cultural ends are realized.

Relationships between cultural rules governing the behavioral and ideological components of social life and the physical environment in which they occur have long been the object of anthropological inquiry. Although much disagreement exists over specific cause–effect relationships between what people think and do and the physical conditions in which they live, authorities generally agree that the constraints and opportunities provided by the physical environment are key factors in the determination of culture.

An example of how physical and social components of culture may be manifest in a specific society (e.g., Trobriand Islanders) is shown in general outline in Figure 11.1. In this case, the following information is essential to understanding the social dynamics of this system, which includes interacting shoreline and inland communities.

1. The society has a matrilineal kinship system. In such a system, the child ("ego" in figure) is taught to trace his allegiance (including mutual obligations) only through his mother's side of the family. All genetic relatives who trace through the same female line constitute a *sib* (or *sibship*). All persons owe their primary obligations to members of their own sib. In the matrilineal sib, ego's mother's brother (MoBr) is equivalent to ego's father (i.e., Fa = MoBr), that is, MoBr is the key reference male and provider for ego.

2. The society is patrilocular, that is, after marriage the female must move to her husband's village. In this particular society, marriageable males and females living in seashore communities are required to find their respective mates in inland communities, and vice versa.

3. Agricultural tasks (in inland communities) and fishing (in seashore communities) are cooperative tasks. Agricultural tasks are performed by the nuclear family (biological parents and offspring). Fishing requires coopera-

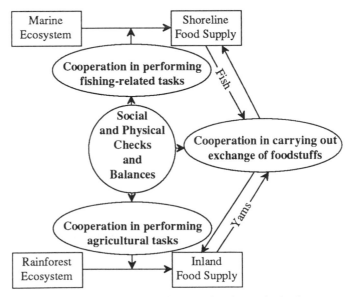

FIGURE 11.1 Overview of food exchange between shoreline and inland communities in an island based matrilineal society.

tion among males who divide their labor among various tasks, including constructing canoes, making nets, paddling, and hauling nets.

As shown in Figure 11.1, the culture provides for a high degree of coordination among the various tasks associated with agriculture (primarily yams) and fishing, as well as for the exchange of foodstuffs between seashore and inland communities. The exchange of foodstuffs is extremely important because it provides a balanced diet in both communities.

A more detailed representation of the cultural mechanisms that insure food production and food sharing is given in Figure 11.2. In this figure, which integrates major features of marine and rainforest food webs with the social system, fish not only serves as a food supply for shoreline communi-

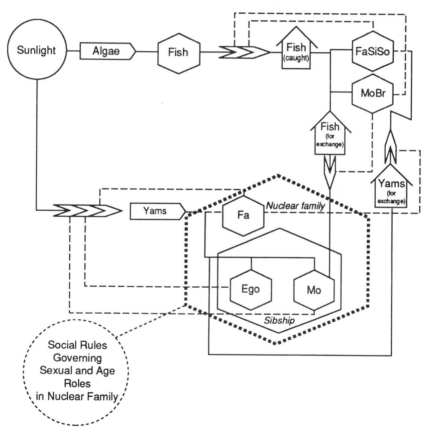

FIGURE 11.2 Examples of energy flow and important kinship relationships among matrilineal island communities. Names of individuals are the English equivalents of those attributed by *ego*. Note: Read *FaSiSo* as "father's sister's son." (See Figure 11.1 for overview of these relationships.) _____ represents flow of energy and/or material; --- represents work performed.

ties but is also the primary means by which the kinship obligations of a brother to his sister's child are met. Precisely the same kinship obligations control the sharing of agricultural produce between inland and shoreline communities.

These kinship obligations, which in the preceding example result in food sharing, are manifest in a variety of rituals that, to an outside observer, might be characterized as religious or legal. However characterized, such kinship obligations permeate all aspects of community life. Therefore, that significant impacts on marine and rainforest food webs might be expected to result in potentially significant social impacts, including impacts on public health (through alteration of diet) and divorce rates, is not surprising.

PERSONAL AND INTERPERSONAL CONTEXT

Just as the two sides of a coin cannot be separated and the integrity of the coin be maintained, so the personal (or individual) side of human existence cannot be cleanly separated from the interpersonal (or group) side. Although a person has individuality, he or she is not independent of others, if only by virtue of the enculturation or socialization process that attempts to program him or her with the collective experience of others. Even when the socialization process apparently fails in directing the thinking and behavior of an individual into preformed patterns, the individual learns by experience of society's attempt, and thus learns (either conformity or nonconformity) through others.

This learning of culture, which takes place in an individual in a social and physical context, should not be seen as primarily a kind of vocational learning that is required to perform a task, such as the learning of a manual skill. More importantly, the enculturation is the learning of prescribed associations between sensory and other informational signals and mental images. For example, the words "cat," "god," and "good" have innumerable meanings, depending on images we have individually learned to associate with them. Surely we are not born with mental images of their meanings. We know what these words mean because of the mental programming of our social experience. In the most general sense, how we **individually translate** the world around us into meanings that are interpersonally shared is the subject of *social perception*.

Role and Status

That we (within a culture) do individually perceive the world around us in the same ways as others (within the same culture) has an important consequence for the organized pattern of social life: the sharing of perceptions promotes social expectations. For example, the perception of "Monday's sunrise" that is shared among many as "the start of a workday" is clearly the basis for those same persons expecting certain things to happen, including traffic jams, the rush for parking spaces, and the arrival and departure of trains and buses according to prescribed schedules. What if "Monday's sunrise" was perceived by bus drivers and conductors as "a holiday" or "a time to change schedule randomly," or by parking-lot attendants as "a day to go fishing"? The pattern or organization of social life in a modern city would be disrupted.

An important concept that is related to the socially shared expectations of behavior is the concept of *role*. The role of an individual is composed of the expectations that others have of that person's behavior because of that person's position (i.e., *status*) within society or some social group (e.g., president, mother, teacher, priest, gang leader). The concept of role is important because it integrates individuals into a matrix of interpersonal expectations of behavior. The contingency of these expectations introduces the normative quality or predictability into social life.

For example, my getting onto an airplane is very much contingent on my perception of the crew (especially the pilot) as being qualified, that is, I expect the crew to act in a certain manner that I have identified, either through direct experience or secondhand knowledge, as indicative of their knowledge of their task. Of course, the crew is unlikely to fulfill its role (i.e., meet my expectations of behavior) if the individuals do not perceive me as a passenger intent on traveling from one place to another. Should I behave suspiciously while undergoing the security check or otherwise be perceived as someone intent on mayhem rather than as a serious passenger, the crew most definitely would not behave ordinarily. By satisfying the mutual expectations of interdependent role players, social perceptions organize social systems and give them the predictability that day-to-day life demands.

Primary and Secondary Groups

One of the elements of social organization that promotes and benefits from the predictability of daily life is the *social group*. Although all social groups may be characterized by (1) some orientation of individual members toward one another, (2) some sharing of values and beliefs, (3) some pattern of relationships among group members, and (4) statuses that are occupied by specific individuals, social scientists recognize a variety of different types of social groups.

For example, some social groups are specifically concerned with the promotion of interpersonal relationships within the group membership. In such groups, the personal identification between members (i.e., *reference*) is of paramount importance. The group may be a criminal gang or a bowling team. Regardless of its specific composition or purported purposes, if group members and actions give priority to the intimacy of interpersonal relationships within the group, social scientists refer to the group as a *primary group*.

In contrast, a *secondary group* is one in which priority is not given to the intimacy of interpersonal relationships, but to some interest held in common by group members. The key characteristic of a secondary group is the formality of its organization with respect to various group activities, including those related to membership application and acceptance, communication within the group, the exercise of power and decision making, and the resolution of conflicts. Typical examples of secondary groups might include a school committee and a working shift in a factory. Although individual members of these groups may expend a great deal of energy and emotion in group activities and on one another, secondary groups are primarily concerned with the achievement of objectives that are common to their memberships and not with providing identity and life meaning to the members.

INSTITUTIONAL CONTEXT

A common-sense view of institutions (which is amusing in the classroom, but misleading in the real world) is that they are names of buildings, for example, legal institutions (as evidenced by courthouses), economic institutions (as evidenced by banks), educational institutions (as evidenced by school buildings), and religious institutions (as evidenced by churches, synagogues, mosques). Approaching the true meaning of the concept of institution by first emphasizing that institutions perform specific functions, such as (1) specifying goals, (2) developing and applying means for achieving those goals, (3) maintaining the integrity of social groups, and (4) resolving disrupting tendencies within social groups, is useful. As a "social construct" that carries on such functions, an institution has been identified as "a complex of bureaucratic organization" or as "rules that govern specific types of social action."

One way to depict an institution is to identify the persons who participate in that institution, their particular functions, their relative positions within the organizational hierarchy, and the roles that they play. Another approach is to identify the rules of behavior and thinking that are inculcated into those who subject themselves to socialization by the institution. These two approaches to the concept of institution are not mutually exclusive. Instead, they complement one another. For example, religious institutions may be seen as hierarchically organized groups of people (e.g., clergy) and as collections of rules (e.g., doctrines, values) that are promulgated within society at large by the functioning of the hierarchy. The combined view of an institution as a bureaucracy and a set of rules yields a more comprehensive and realistic appreciation of an institution as a functioning societal entity than does either view alone.

As collections of rules for governing overt behavior and ideology, different social institutions may promulgate similar behavior and values. Thus, the rule "Thou shalt not kill" may be found as a basic rule in the socialization efforts of religious as well as legal institutions. Of course, the socialization undertaken by one institution can also conflict with the socialization undertaken by another. For instance, the religious code of behavior exemplified in the rule "Do unto others as you would have them do unto you" can be expected to be in direct conflict with a variety of behaviors promulgated by those economic institutions historically based on monopolistic competition.

In general, the consistency of rules promulgated by different institutions varies with the general type of society. One type of society has been variously described as "pluralistic" or "associational," or as having "organic solidarity." A *pluralistic society* is typically large and depends on numerous specialized and diverse (even conflicting) groups and interests. In pluralistic societies, rules (behavioral and ideological) promulgated by different institutions are often inconsistent with or contradictory to one another. In contrast, a *monistic society* (*communal society;* a society having "mechanical solidarity") is typically small, physically isolated, physically homogeneous, and largely governed by unwritten traditions. In such societies, rules promulgated by diverse social institutions tend to be mutually congruent or at least mutually supportive.

The large size and internal diversity and complexity of contemporary pluralistic societies present historically unique challenges regarding the role of traditional institutions as the primary "programmers" of the "human computer." Whereas kinship groups were previously important institutionalized means of receiving and processing information about the outside world, other institutions have now become important, including governmental and educational institutions. Because of the coexistence of diverse institutions that have overlapping jurisdictions but different values and interests, pluralistic societies have a high potential for creating confusion among people with respect to what they should do and think, and why. This condition historically has been referred to as *anomie.*

Although the concept of anomie has been used to refer to the normlessness of pluralistic society (a condition often described as *social disorganization*), more modern usage treats anomie also as a property of individuals (called *personal disorganization*) who must confront a confusing and even self-contradictory structure. This approach reflects the ongoing trend in social science to integrate institutional, personal, and interpersonal components of social systems.

IMPACT ASSESSMENT AND SOCIAL DYNAMICS

When faced with the task of describing a human being, perhaps the easiest way to proceed is to begin with a simple physical description. If this is not sufficient, one can discuss behavioral and even mental attributes. With respect to the assessment of impacts on the social environment, descriptions of the social being of humankind have too often been limited to the elementary level of gross social anatomy, and have rarely progressed to the level of social dynamics.

For example, many assessment teams might view the social being of humankind to be essentially composed of bodies, buildings, and money, that is, the numbers of persons to be relocated or otherwise inconvenienced, buildings to be razed or constructed, and money to be spent and generated by project development. However, the social being of humankind is actually infinitely more complex than the mere numbers of bodies or the physical and economic artifacts of human life.

The social being of humankind is built of individual emotions as well as on group aspirations and institutional goals. It is built of values and beliefs, of human expectations of what will happen in the future, as well as memories of previous cultural experience. It is built of the diversity of interests as well as of consensus. Of course, one cannot easily measure beliefs, count aspirations, or give numerical values to interpersonal relationships. Nevertheless, these aspects are real, and the mechanisms by which these largely unquantifiable elements of social life are integrated and actually expressed in the patterns of social life constitute social dynamics.

Certainly good reasons explain why assessment teams have generally overlooked the impacts of projects on personal, interpersonal, and institutional dynamics and have focused instead on project impacts on social artifacts:

- The evaluation of social impacts implies purposeful "social engineering," a particularly politically sensitive issue for many decision makers.
- Assessment teams have most often been dominated by individuals trained primarily in the engineering and natural science disciplines.
- Social sciences are widely viewed as a potpourri of unsubstantiated opinions.

With respect to these facts, we must identify some basic principles that should underlie each assessment of project impacts on the social environment.

First, the assessment team should understand that **all projects—regardless of their stated purpose, their location, or their design—do accomplish some degree of "social engineering."** All projects maintain or change, enhance or decrease, introduce or remove, restrict or distribute, direct or redirect, or otherwise affect patterns of human life. That any proposed project could ever be undertaken if assessors could not demonstrate that the project would sometime affect someone somewhere is absolutely inconceivable. Because projects must somehow affect social life if they are to be undertaken at all, essentially only two types of projects exist—those that result in foreseen consequences and those that result in unforeseen consequences. Thus, the problem of assessing impacts on the dynamics of social systems is not that projects involve social engineering, but that the assessment process can fail to identify or consider relationships between proposed projects and social consequences.

Second, the assessment should include individuals who, by their training or their experience, are sensitive to the concerns, concepts, and analytical techniques of professional social scientists. Certainly, the assessment team need not include professional level biologists and ecologists as long as the team has access to such professionals and to the relevant literature. The assessment team need not include professional level sociologists, social psychologists, or political scientists, as long as the team has access to these professionals and their literature. However, access alone is not sufficient. The access must be used by team members who are both capable and serious in pursuance of information and data pertaining to social dynamics.

Too often, individuals who are experientially qualified to gather and process data and information on the social environment have been excluded from the assessment process. For example, right-of-way professionals are those personnel who undertake the appraisal and acquisition of real property for project development. Much of the success of right-of-way professionals in their work can be attributed to their sensitivity to people, social issues, and concerns and to their ability to communicate with people with diverse backgrounds, interests, and training. However, despite these qualities and skills, right-of-way personnel are typically excluded from assessment teams. This particular situation, which has persisted in the United States for over 20 years, is likely to reflect the stubborn bureaucratic tendency to view impact assessment primarily as a task for engineers or personnel trained in the physical sciences.

Third, the assessment team must disregard its own preconceptions of the various social science disciplines and deal with social scientists and their literature the same way it deals with natural scientists and their literature.

SOCIAL DISTRIBUTION OF IMPACTS

Essentially ignored during the formative period of impact assessment, the manner in which environmental impacts are actually distributed within the social system is of increasing concern. Impacts on public health (Chapter 13) as well as economic impacts (Chapter 14) are particularly amenable to distributional analysis.

As discussed in Chapter 6, an important input into the valuation of impacts on the social environment is the issue of social equity. This issue, often referred to in the United States as *environmental equity,* is predicated by the concern for the equitable distribution of environmental risks and benefits within a population characterized by a diversity in race and income. Other demographic factors that might be included in a consideration of social equity include sex, age, geographic location, and occupation. In some societies, social class, religious faith, and other socially significant statuses (e.g., indigenous or native peoples, ethnic groups) may be important dimensions of social equity.

Of course, no issue is more politically or culturally sensitive than the relative social distribution of environmental risks and benefits. Although increasingly recognized as an important component of impact assessment in some legal jurisdictions, a consideration of the distributive dimensions of environmental impacts is prohibited in (or effectively avoided by) others. To the degree that the distributional dimensions of environmental impact assessment are ignored, the assessment of social impacts becomes increasingly an academic exercise and less and less a practical tool for an informed decision-making process in project development.

CHAPTER 12

PERSONAL AND INTERPERSONAL IMPACTS

Impacts on public health and safety and economic impacts are addressed in more detail in Chapters 13 and 14, respectively. In this chapter, the focus is on how projects may otherwise affect social components and dynamics. Health, safety, and economic aspects are discussed only to the extent necessary to establish linkages between components and dynamics.

A primary consideration in any assessment of social impacts is that personal, interpersonal, and institutional components of social life are strongly integrated. Thus, a project-related impact on a component may, in fact, rapidly resonate throughout a nexus of personal, interpersonal, and institutional dimensions, gaining new content and new direction and becoming manifest in a wide variety of consequences. Although such systems-level effects also characterize highly integrated biotic, abiotic, and ecological dimensions of the physical environment, they are particularly important in social systems because the flow of information between personal, interpersonal, and institutional components is infinitely more diverse, rapid, and circuitous than the flow of matter and energy that connects and informs the various components of the physical environment. Moreover, the perceptual, cognitive, intellectual, emotional and behavioral repertoire of social dynamics would appear to be limitless—always guided by but not necessarily bound by cultural imperatives and always informed by but not necessarily confined by previous experience.

For the purposes of impact assessment, distinguishing two sources of social impacts—project-level and action-level—is useful.

Project-level sources are social perceptions of the proposed project rather than actual action or activity undertaken as part of project development. At this level, the potential for impacts arises out of the emotions triggered in individuals who perceive the proposed project as an intrusion into their lives. In some instances, the intrusion may be specifically perceived as a physical threat (e.g., a nuclear power plant). In others, the intrusion may be more generally perceived as a nuisance, an unwarranted assault on privacy, or even a flagrant reminder of social, economic, or ethnic prejudice.

Action-level sources are objectively definable project-related actions, activities, and design features and attributes. At this level, which tends to receive the vast majority of the assessment effort, the focus is on the causal chains initiated by actual project development involving manipulations of matter and energy.

The importance in distinguishing between these two levels of impact is that, unlike the physical environment in which impacts must arise out of changes in the objective attributes of matter and energy, the social environment is subject to impacts arising out of the subjective workings of the human mind, a fact too often ignored in the environmental assessment process.

SOME EXAMPLES OF PERSONAL AND INTERPERSONAL IMPACTS

Certainly many types of project result in an infusion into ongoing communities of different peoples with different life-styles, values, and beliefs. Such ideological and behavioral differences may represent alternative choices to the values and aspirations that have long found their expression in the controlled socialization of the young by local institutions.

The consequence of conflict between historical and new alternative socializations lies not only in the potential for overt clashes between opposing groups, but also in the minds of individuals who, faced with alternatives not previously available, find their traditional roles challenged. In opting for new roles, these individuals can experience *role conflict* (i.e., the situation in which one attempts to meet conflicting expectations of others who are important to the concept of self) and must behaviorally and/or psychologically adopt techniques for dealing with the contradictory expectations of others. At the personal level, the confusion inherent in significant role conflict may be described as a condition of personal disorganization. When substantial numbers of persons within a society undergo personal disorganization, the society is experiencing social disorganization.

Note that no absolute sociological value is attributed to role stress and conflict or to their absence, nor is any absolute value attributed to social change or to the maintenance of the *status quo*, nor is any absolute sociological value attributed to personal and social disorganization or to personal and social organization. The positivity or negativity of each of these conditions must be determined within the context of a real and ongoing social system. This context, after all, can be determined by what the people who are actually involved think about social change, about their aspirations, and about their frustrations and fears. Of course, more objective measures of the consequences of personal and social disorganization may be attempted in terms of certain social disfunctions including divorce, suicide, substance abuse, and aggression.

Location Phase of Project Development

Projects that require long-term planning and construction (several years) in relatively isolated areas (e.g., oil pipeline project in Alaska, highway projects in central Brazil) typically require the import of a large number of skilled workers into the project area. Certainly no other situation is more likely to result in social conflict than the mixture of local populations and well-paid strangers in an isolated area over an extended but finite period.

Example 12.1 A local economic boom is likely to result in a relatively high rate of inflation that is not likely to be prohibitive to the well-paid project worker, but may quickly become prohibitive to the "breadwinner" who is otherwise employed in the project area but whose income remains fixed (e.g., a teacher, government employee, subsistence farmer). The resulting personal and familial stress among such affected persons may become manifest in a variety of social dysfunctions, including alcoholism, familial abuse, and divorce.

Example 12.2 Significantly higher paying project-related jobs may not only recruit local workers away from their previous work but may also (and probably more importantly) serve as the functional model of an alternative life-style that directly contradicts local values and beliefs. Thus, a socialization process that gives high value to the importance of education becomes seriously suspect when confronted by the fact that money, adventure, and pleasure are more immediately and plentifully available through physical labor.

Example 12.3 Redirected values and life-styles include not only those directly related to project-defined jobs but also those related to project-related opportunities, such as prostitution, drugs, and other criminal activities, each of which has important ramifications regarding public health and public safety that may easily persist beyond the completion date of the project.

Example 12.4 Alternative life-styles and values exhibited by imported workers may, of course, repel as well as attract local residents. These life-styles may win new converts or reinforce historical barriers that effectively separate people, including differences in religion, ethnicity, and race as well as differences in behavior regarding the diverse necessities of daily life. When such differences are perceived by local residents as real threats, blasphemies, or insults, violence is likely to arise.

Of course, most projects requiring impact assessment do not require the import of large numbers of strangers who carry significantly different life-styles and values into relatively isolated communities. Most projects involve a workforce that is present in the project area for relatively short periods of time and is essentially indistinguishable from relatively local populations. Under such pluralistic conditions, local perceptions of the proposed project are most heavily influenced by past and ongoing patterns of community self-image.

Example 12.5   The persistent location of certain types of projects typically associated with high risk (e.g., hazardous waste treatment plants), filth (sewage and other waste disposal plants, landfills), or high levels of air emissions (e.g., highways) in areas typically populated by the poor and ethnic or racial minorities rightfully raises the question of social equity, especially when the primary economic benefits of the project are realized by a more politically and economically more powerful, far-removed population.

The disproportionate distribution of risks (whether actual or perceived) or generally acknowledged unpleasant, noxious environmental attributes in a manner consistent with historical social and economic bias and division can only poison the diverse political and socialization processes of a pluralistic society.

Example 12.6   The location of many projects that are not typically associated with high health or safety risks but nonetheless can be associated with persistent nuisance (e.g., noise associated with traffic, low and high frequency electronic noise associated with transformers, traffic congestion associated with business development) may also establish social inequities within and among communities. In many instances, the perceived nuisance may not be the result of changes in environs that can be measured objectively in terms of decibels of sound or density of traffic, but may be defined subjectively by persons and groups, for example, architectural features of project facilities that do not conform to neighborhood patterns or to the perceived aesthetic quality of project facilities.

The location of specific facilities related to project development implies the dislocation or absolute loss of other community features, including previous landowners, structures, or open areas. Impacts resulting from such dislocations depend on the social roles and functions these community components may perform.

Example 12.7   Regardless of the personal impact on dislocated persons and families, social impacts include interpersonal impacts that may arise from the removal of individuals from an ongoing social network. Depending on the attitudes, values, and behaviors of those persons to be displaced and of the community as a whole, the *de facto* removal of certain individuals from a locality may be viewed by the community, in some instances, as a positive reinforcement of shared community values and, in others, as quite the opposite.

Example 12.8   The loss of structures and open areas within a community as a result of project development may also lead to a variety of secondary impacts, depending on the contribution of those structures and areas to both socially acceptable and deviant values and behaviors.

Regardless of its primary use as a school, a place of religious worship, a governmental office, or even a privately owned business, a building may also serve an important community function as a place of public assembly important for recreational and other neighborhood social functions. Although the relevant educational, religious, governmental, and business functions may be easily relocated without functional interruption, ancillary but no less socially significant neighborhood functions may not be as easily or successfully relocated. Of course, buildings and their associated properties as well as open areas may also be the site of behavior that is contrary to local community values. Thus, the loss of such structures through project development may reasonably be considered a positive reinforcement of those values.

Construction Phase of Project Development

Potential impacts during the construction phase of project impact are typically those related to health and safety, including;

- vehicular accidents, due to enhanced traffic congestion in work areas
- respiratory ailments due to fugitive dusts and/or the exhausts of heavy construction equipment
- on-site storage of hazardous chemicals and materials that may escape into the community as surface runoff, groundwater contamination, or air contamination
- improper fencing or isolation of potentially dangerous work areas

Other types of impact depend on site-specific aspects of the community in which the work is performed.

Example 12.9   Many types of construction activities produce substantial noise and vibration, including such activities as blasting, drilling, pile-driving, and the operation of heavy equipment, such as vehicles, jackhammers, and cranes. In most instances, such construction activities are undertaken with little specific awareness of the consequences of such noise and vibration on particularly sensitive individuals within the surrounding community, such as bed-ridden patients receiving home care.

Although public notice of particularly significant construction-related noise (e.g., blasting) is typically given in the local mass media, information is rarely given or received on a residence-by-residence basis. Thus, the possible personal concerns of individuals are largely ignored.

Example 12.10   Any construction in a residential area is essentially an intrusion into the day-to-day life of the community. Although activities specifically related to the act of constructing may be tolerated as merely temporary inconveniences, inappropriate or ill-advised behavior of construction crews is not rationalized as easily. Examples of inappropriate behavior that may be considered innocent by the construction crew, but is viewed much more seriously by individuals and groups within the community, include the subjection of citizens to visual and verbal sexual innuendo, raucous breaktime noise during local religious holidays or periods of personal bereavement, and conversations among construction workers at local lunch counters that, by content or tone, belittle, insult, ridicule, or otherwise disregard local values and customs.

Operational and Maintenance Phase of Project Development

Largely underemphasized in historical approaches to impact assessment, which have focused on the construction phase, the operational and maintenance phases of project development are typically the phases in which major impacts become manifest. Although project engineers consider a project to be completed once it has been constructed, precisely at this time a new environmental dynamic has become established with consequences that are not necessarily predictable on the basis of the goals and objectives of project design. Extreme examples include roadside rest areas that become target areas for robbery and murder, and the opening of virgin timber and mining resources that too often heralds the decimation of indigenous peoples.

Example 12.11   In-place design features of many projects may often function as *attractive nuisances*, with socially important consequences. For example, isolated or poorly lit parking lots may become assembly points for persons engaged in raucous or other types of undesirable behavior. Parking lots as well as walkways might be used for recreational purposes (e.g., rollerskating) that are contrary to their intended use and enhance the probability of vehicular accident. Retaining walls and other landscaping features, such as fountains and reflecting pools, might also be misused and contribute to serious accidents.

Example 12.12   The operation of extensive developments, such as malls, often results in the careless disposal of foods in parking lots which may attract scavenger birds such as gulls. Depending on the density of such populations, these gulls may become of concern regarding vehicular safety and, in areas dependent on surface water supplies for potable water, the contamination of those supplies by bird droppings. Other scavenger species may also be of consequence to public health.

Example 12.13   Some projects offer particularly useful opportunities for public education that are typically not envisioned during the course of project design. Such opportunities might include tours of facilities (e.g., pumped storage power plants, transportation complexes, sewage and water treatment plants) arranged for primary and secondary school classes, as well as lectures and demonstrations for the general public. Depending on the nature of the project or facility, recreational opportunities may also be presented, including nature walks in the protected watershed of a water resource management facility.

Example 12.14   The staffing of operational projects and facilities typically mirrors the power structure of the larger society and, in societies in which the distribution of power reflects deeply engrained socioeconomic inequities, such projects and facilities emphasize larger social inequities at the local level. When local underprivileged populations typically occupy lower status positions in the completed local facility and nonlocal representatives of the socially privileged population occupy the higher status positions, a likely result is the exacerbation of social conflict at the local level.

Example 12.15   Maintenance activities associated with an operational project or facility, such as the control of nuisance vegetation (e.g., application of herbicides, selective cutting), waste disposal (e.g., snow, dirt, trash, hazardous waste), and structural maintenance (e.g., painting, application of preservatives, sandblasting), are typically conducted according to a schedule primarily based on facility operations. Potential effects on the surrounding community are often overlooked.

For example, the application of herbicides (or other pesticides) and spray painting may result in fugitive toxic aerosols escaping to surrounding environs, with possible public health implications. Fugitive chemicals and dusts may also settle out on personal property, resulting in actual damage or nuisance. The on-site storage of snow, dirt, and trash not only presents a possible safety hazard, especially for children, but may also conflict with community aesthetics.

NEED FOR PUBLIC PARTICIPATION IN THE ASSESSMENT PROCESS

As indicated by many of the examples already given, social impacts include not only objective events and situations, such as the displacement of a certain number of families, but also subjective realities as defined by the individuals who experience them. To omit the subjective dimension of human life from the impact assessment process is a failure to consider real people. How can the assessment team have any assurance that it is appropriately considering the real concerns of real people?

Wherever practiced, environmental impact assessment has typically involved the public as part of the decision-making process of project development. Although the specific nature and degree of public participation varies with the legal jurisdiction of varied assessment rules and regulations, public participation may generally be viewed as predicated on specific executive, legislative, or judicial definitions of various rights of the public, including the *right of access to information* gathered during the assessment process, the *right to contribute information* to the assessment process, and the *right to challenge decisions* made in the progress of or in light of the assessment effort. Unfortunately, general policy statements concerning the importance and value of public involvement in the assessment process do not necessarily reflect or influence actual events in an assessment project. With respect to the thousands of assessments that have been completed, with the notable exception of the Canadian experience, public involvement has largely remained an unfulfilled ideal.

Regardless of procedural compliance with the objective of public involvement, the fact remains that **assessment teams typically view the public more as an adversary than as a partner in the assessment process.** This perspective precludes any serious consideration of the constructive role that the public may play in the identification and evaluation of potential impacts on environmental components and dynamics.

Instead of concentrating on the risks and difficulties of including the public in the assessment process in a meaningful way, assessment teams must concentrate on the benefits that can be derived from an enhanced and much expanded communication among team members and the public. Communication between the assessment team and the public is the key to public participation. However, communication is more than a simple transmission of information from one group to another; it is an active and constructive **exchange of information, meaning, and opinions.**

The importance of communication as a means of achieving constructive public involvement in the identification and assessment of environmental impacts is being increasingly recognized. As early as 1976, this priority was formalized by the United States Federal Highway Administration in the following dicta that all assessment teams are well advised to follow:

- communicate with the public as early as possible
- communicate with as many people as possible
- communicate through as many different means as possible

Overview of Roles of the Public

Within the broad scope of assessment goals and objectives, the public can constructively participate in each of the following tasks. Communication between the assessment team and the public must be undertaken to facilitate the public's peformance of these functions.

1. Provide data and information that is essential for the assessment of impacts on the physical and the social environment.

2. Help identify local citizens and groups with special expertise that might be used by the assessment team for specific tasks.

3. Identify local and regional issues that should be addressed in the assessment process.

4. Provide historical perspective to current environmental conditions and trends in the local and regional area of proposed project development.

5. Help generate field data.

6. Help provide "field-truth" for key data generated during assessment.

7. Provide criteria for evaluating the significance of identified impacts.

8. Identify project alternatives.

9. Suggest and help organize forums and mechanisms for public participation in the assessment process.

10. Monitor the relevance and adequacy of ongoing assessment efforts.

11. Review interim assessment reports and findings for public readability and relevance to local issues and concerns.

12. Help analyze and evaluate direct, indirect, and cumulative impacts of project development.

13. Help define the scope of work and schedule for the overall assessment process.

14. Provide liaison between assessment team members and key organizations and other public groups and individuals.

15. Review, comment on, and make recommendations with respect to the qualifications and planned approach of consultants to be used by the assessment team.

16. Identify and evaluate potential mitigation measures that might be incorporated into project design or management.

Clearly, for the assessment team to presume that any one public group or organization can or would desire to undertake all these tasks is unreasonable. Different groups and individuals in a local area may undertake one, several, or none of these tasks, depending on individual interests, availability of time, knowledge, and experience. Therefore, the responsibility of the assessment team is to help diverse public groups and individuals identify assessment functions that each may undertake successfully. This responsibility should be given the highest priority by the assessment team. In fact, actions undertaken to fulfill this responsibility should be among the very first actions initiated by the assessment team.

Public Participation Techniques

A wide diversity of techniques has been used to facilitate public participation in impact assessment. These techniques, which have been variously described as media-based, research, political, structured-group, large-group, bureaucratic-decentralization, and intervener techniques, include (but are not limited by):

Media Techniques
Participatory radio and television
Newsletters
News releases
Public service announcements
Paid advertisement

Research Techniques
Sample polls
Community profiles

Political Techniques
Citizen referendum
Lobbying

Structured-Group Techniques
Nominal groups
Delphi techniques
Charrette

Workshop
"Show-me" trips

Large-Group Meetings
Public hearings
Public meetings

Bureaucratic Decentralization
Field office
Information van

Interveners
Ombudsman
Citizen advisory committees
Advocacy planning
Ad hoc committees

Note that the various techniques for facilitating public participation in environmental decision making have often been applied with specific reference to target groups, including (1) the general public, (2) public officials, and (3) members of the general public with professional expertise. Most often, the emphasis has been given to public officials and professionals rather than to the general public, suggesting that much room for improvement exists in designing and implementing public participation.

In selecting and implementing any technique for involving the public in the assessment process, the assessment team should assure itself that adequate consideration has been given to the following questions.

1. What specific assessment objectives can be achieved by the proposed technique?

2. What are key criteria (e.g., physical setting, timing, nature of target group) for the successful utilization of the proposed technique?

3. What follow-up actions and related budgetary, personnel, and information resources will be required if the proposed technique is implemented?

4. How can the implementation of the proposed technique be monitored to insure the timely correction of any counterproductive conditions and tendencies?

5. How can local conditions (e.g., attitudes, previous experience with public participation measures) influence the successful utilization of the proposed technique?

6. What criteria can be used to insure the most appropriate assignment of team personnel with respect to the successful utilization of the proposed technique?

The appropriateness of any type of public participation program, and the particular skills required for the successful implementation of that program, are largely determined by local conditions, including local attitudes and concerns, as well as by the previous experience of citizens with public participation programs. Thus, any recommendations for utilizing a particular technique (or combination of techniques) to involve public groups and individuals in the assessment process must be made in light of the team's knowledge of these location conditions.

As a general rule, the potential for initiating counterproductive programs of public involvement is greatly increased whenever assessment teams fail to define site-specific factors. In short, any failure of a public participation program must be attributed to the shortcomings of the assessment team and not to the public. Any other approach will invite an infinite number of self-serving excuses for why the public should be excluded from environmental decision making.

The attitudes of members of the assessment team concerning public participation cannot be overemphasized. Team attitudes are key factors in implementing a successful public participation program. The idea that the assessment team is doing the public a favor by dealing with the public at all, or the idea that the assessment team is more knowledgeable of, is more interested in, or has more of an investment in the total environmental ramifications of project development than the public itself has no place in impact assessment. Any public that rightly perceives these attitudes among assessment team members is fully justified in assuming that the public is being used rather than served.

CHAPTER 13

IMPACTS ON PUBLIC HEALTH AND SAFETY

The identification and assessment of impacts on public health and safety require the integration of physical, chemical, biological, social, and psychological factors related to project development. As a general rule, the more comprehensive the assesment of project impacts on the physical and social environment, the more comprehensive the assessment of project impacts on public health and safety, including impacts on (1) physical well-being (i.e., safety), (2) physiological well-being (i.e., health), and (3) psychological well-being (i.e., welfare).

Project impacts on each of these dimensions of human well-being may occur throughout project development and may be influenced by changing environmental conditions that, even if not related to project activities, can nonetheless potentiate project impacts on human health. For example, a particular project may result in the long-term buildup of potentially stressful chemicals in drinking water supplies. Although project concentrations may be well within public health standards for potable water, a long-term trend in the public's dietary intake of those same chemicals may also be created. Certainly the total dietary intake of potentially stressful chemicals is not a direct or even an indirect result of any single project. However, incremental contributions from a particular project can exacerbate the problem in a local area.

This example demonstrates a particularly important point about the assessment of project impacts on public health: **the assessment team must evaluate the proposed project's impacts on public health in light of the total environmental context in which people live.** In keeping with such a comprehensive perspective of public health, the assessment team must rid itself of a rather depressingly common attitude that has often been pervasive in assessment teams more interested in getting a project built than in doing a comprehensive assessment. This attitude is exemplified by the following remark: "The safety risks of this project are less than the risk people take when they get into their automobiles to drive to the store for a loaf of bread. They don't seem to worry about the risk, so why should we worry about the so-called risks of this project?" **The business of the assessment team is not to decide which public health risks people *should* take. Instead, the assessment team should identify the risks that people *will be faced with* as a result of project development.**

Public health considerations of project development are often restricted by a historical perspective of health as the sole disciplinary domain of the natural sciences. According to this perspective, threats to human health can be fully described in terms of microbes, chemicals, and those physiological aberrations that are common to humans. Certainly this perspective permits consideration of mind, emotions, or the intricacies of social organization as important or even relevant factors of human health. Thus, impact assessments tend to deal with the categories of biological and ecological impacts, social impacts, economic impacts, aesthetic impacts, and public health impacts as separate and independent dimensions of human existence.

However, any perspective of public health that does not accommodate social and behavioral factors as well as physical, chemical, and biological factors is outdated and has no place in contemporary impact assessment. This premise was succinctly stated in two precepts presented by the United States Department of Health, Education, and Welfare over 20 years ago.

- People's reactions to the people around them and to the social groups of which they are members may have a major influence on any disease process.
- The effect of the environment on illness cannot be fully understood unless people are considered within the context of the social groups of which they are a part and unless their perceptions of their environment are also considered.

In view of these considerations, individuals who are responsible for assessing public health impacts learn to deal with a real world in which physical and social phenomena actually interrelate. Such a world is certainly more confusing than a neatly compartmentalized-by-discipline world, but is also the world in which real people live.

A SYSTEMS APPROACH

A consistent theme throughout previous chapters has been the need for the assessment process to take account of the dynamic interrelationships among diverse environmental components and processes. Accordingly, the approach to assessment has been described in this volume as a systems approach that focuses on causal chains that dynamically link potential causes and subsequent effects, regardless of the disciplinary properties of either causes or effects. A systems approach to public health not only helps achieve a comprehensive understanding of project impacts on public health, but also helps achieve the integration of public health impacts with other impacts on the physical and social environments that must be considered by decision makers. Although no theoretical necessity for it exists, practical experience with assessment suggests that a simple checklist approach to the assessment of impacts on public health is a good indication that the overall assessment of project impacts will be poorly integrated and will, accordingly, fail to identify important indirect impacts of project development and operation.

Basic components of a systems approach that have long been useful for the assessment of impacts on public health are included in Figure 13.1 and may be summarized:

1. **Environmental factors:** Include physical, chemical, biological, and psychosocial factors; these factors (e.g., pathogenic bacteria, toxic chemicals, metallic dusts), as well as their temporal and material variations within the environment, provide the source of public health insult or risk
2. **Environmental media:** Act as the vehicles by which environmental factors are carried through the various components of the physical environment
3. **Human activities:** Determine human experience with and exposure to environmental media and environmental factors
4. **Moderating factors:** Qualitatively and/or quantitatively influence the human response to environmental factors

Various typologies of human "diseases" in current use testify to the importance of the systems overview of public health depicted in Figure 13.1. For example, the incidence of "pollutional diseases," including emphysema, bronchitis, and various types of cancers, is well known to increase with increasing exposure to specific environmental contaminants (i.e., environmental factors) via the workplace atmosphere. Genetic constitution and the general health and life-style (e.g., smoking) of exposed persons also play important roles in determining the actual risk of exposure to such contaminants. Similarly, allergies may be triggered by a wide range of chemicals present in the workplace and the home, and may be mediated by pre-existing health conditions.

With respect to any characterization of "disease," we must understand that a one cause–one disease concept is inappropriate in any environmental overview of public health. Any disease may have more than one causal antecedent. Moreover, human exposure to a known cause of a disease

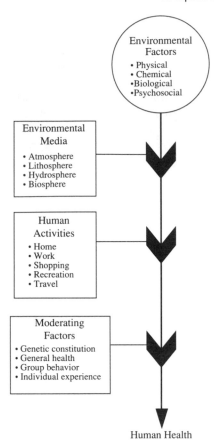

FIGURE 13.1   A comprehensive overview of key factors in determining environmental effects on human health.

does not necessarily result in the manifestation of that disease, nor does knowing a cause of disease necessarily mean that we can prevent the disease from occurring.

The complexity of the multiple-factor etiology of human disease cannot be overemphasized. This complexity reflects the interrelationships among physical, chemical, biological, social, and psychological components and dynamics of the total human environment. Incomplete knowledge and understanding of these interrelationships lead to disagreement over the most important factor for any particular public health risk. For example, although physicians can usually control hypertension with drugs or diet, and although they have investigated various physiological, environmental, and sociological factors, they still do not understand precisely what causes ~90% of the cases of hypertension.

The complexity of public health issues and the existence of diagreements among experts about the causes and contributing factors of health risks are properly given emphasis here because of the misuse that assessment teams can make of these two facts. For example, an assessment team may be tempted to ignore a certain public health risk merely because experts disagree on the significance of that risk. Thus, the team can reply, "If the experts can't agree, how can we, who are not experts, evaluate the risk?"

This question can have only one response: that neither the complexity of environmental components and dynamics nor our imprecise knowledge of them can justify any failure to consider those components and dynamics in the decision-making process of project development. With respect to the assessment of public health impacts, this general guideline for impact assessment may be translated into the following rule: **Any assessment of project impacts on public health that does not take into account the total environmental context of local and regional populations is inadequate by any measure of contemporary environmental health science.**

EXAMPLES

Although projects may contribute directly to the health risks of the public (e.g., industrial contamination of air and water with potential carcinogens), projects may also contribute indirectly (but not necessarily negligibly) to the health profiles of local and regional populations. A systems overview of public health is therefore necessary to identify these indirect impacts.

Example 13.1   As discussed in Chapter 7, a variety of project-related activities (e.g., excavation, rapid fertilization) can result in the eutrophication of lakes or ponds. In this sense, eutrophication might be considered a direct impact of project development.

Some evidence suggests that submerged aquatic plants that characterize many eutrophic lakes promote swimmer's itch (schistosome dermatitis) by providing habitat to snail species that harbor the causative organism. In this sense, the increased incidence of swimmer's itch might be considered one of various indirect impacts of project development. Aquatic macrophytes also provide habitat and food for other animal vectors of human diseases, such as malaria, filaria, and fascioliasis.

Dense populations of blue-green algae (Cyanophyta), which are also associated with eutrophic ponds and lakes, have also been associated with human health disorders, including contact-type dermatitis, symptoms of hay fever, headaches, nausea, various gastrointestinal disorders, respiratory disorders, and eye inflammation.

Example 13.2   If a pond becomes eutrophic primarily because of project-related inputs into the water such as septic and industrial runoff, then additional health risks derive from the specific nature of those contaminants including bacteria, viruses, heavy metals, pesticides, and various organic chemicals. Under certain conditions, some chemicals can concentrate in fish or other aquatic organisms that may be consumed by humans (see "Bioaccumulation" and "Trophic Magnification," Chapter 7).

As these examples indicate, any realistic appraisal of the public health implications of project-mediated eutrophication requires consideration of a variety of factors, including:

- the specific ways in which the project enhances eutrophication (i.e., toxic, nontoxic, and septic contamination),
- the specific uses of water resources undergoing an enhanced eutrophication (i.e., primary or secondary recreation, public water supply),
- aquatic habitat made available (through eutrophication) to disease-bearing organisms, and
- regionally identifiable disease vectors (freshwater snails; malarial parasite; etc.).

Of course, the most important consideration in the preceding examples is whether or not eutrophication might have implications for public health.

Unless individuals responsible for the assessment of public health impacts take a total environmental perspective of public health, issues such as eutrophication are likely to be left to the consideration of individuals principally interested in fish, water quality, and aquatic ecology. In other words, the proof of an assessment team's total environmental perspective of public health is in the continual exchange of information between individuals with primary responsibility for assessing public health impacts and individuals with primary responsibility for assessing other types of impacts, including impacts on the physical and social environment. The simplest rule to follow is that **each project impact, whether on the physical or the social environment, must be evaluated for its direct and indirect influence on public health and well-being.** This objective cannot be accomplished if public health is considered a sesparate issue that can best be handled by people who are isolated and insulated from the overall assessment process.

In considering the public health ramifications of all project impacts on the physical and social environment, the assessment team must understand that "positive" impacts on one component of the environment can nevertheless be (or lead to) "adverse" impacts on public health.

Example 13.3   The conversion of construction borrow areas to new wetland resources, for example, can enhance regional habitat for waterfowl.

In fact, this technique has long been practiced as an important mitigation measure. However, the same wetlands can result in health and safety hazards for humans. These regions may serve as habitat to disease-bearing insects, rodents, and other mammals; they can attract waterfowl that, once in the local area, may contribute significant amounts of fecal matter to surface supplies of potable water; they can also provide hazards to children. Moreover, depending on the chemical constituents of runoff water received by the wetland, and the location of the wetland with regard to groundwater aquifers, such wetlands may also become the means by which groundwater aquifers used for potable supplies become contaminated with toxic chemicals.

All too often, assessment teams fail to consider the full range of public health consequences of "positive impacts" of project development. As the preceding examples suggest, an exclusive concern for wildlife can lead to serious public health risks. Although no objective technique exists for balancing physical and social "pluses" with public health "minuses," the assessment team has the responsibility to assess all project impacts in light of all their potential consequences to public health and well-being.

A particularly troubling tendency in impact assessment is to overlook potentially adverse health impacts that may often be directly associated with project-mediated economic gains such as the expansion of the local job market.

Example 13.4   While hypertension, genetic constitution, cigarette smoking, and diabetes are well-known risk factors in the development of cardiovascular disease, these factors alone are considered to explain only about half of all myocardial infarctions. Other significant risk factors include occupational exposures to chemicals such as carbon disulfide, carbon monoxide, and nitrates.

A project-mediated expansion of jobs for local workers such as electroplaters, oil processors, boiler room workers, diesel engine operators, and explosives makers, who experience relatively high occupational exposures to carbon disulfide, carbon monoxide, or nitrates, is therefore also an expansion of the local population that is at increased risk for cardiovascular disease.

Project-mediated job expansion may not only adversely affect the health profile of a local population but may also contribute significantly to the disproportionate distribution of disease and disability within that population according to sex, race and ethnicity, socioeconomic status, and age.

Example 13.5   When the project-mediated job expansion is primarily realized in the growth of work typically performed (on the basis of social custom) by women (e.g., clerical work, assembly of electronic equipment), women are more frequently exposed to certain chemicals than men are. Examples of such chemicals include ozone, which is associated with the operation of many office photocopiers as well as with the high altitude ventilation of airplane passenger cabins; trichloroethylene, which is associated with the assembly of many electronic devices; and methanol, which is associated with the operation of many duplicating machines. Ozone is a respiratory irritant and cardiac depressant. Methanol can cause severe visual impairment. Trichloroethylene causes cancer in animals and may damage the liver, kidney, and central nervous system.

Of course, occupational exposure to certain hazardous chemicals and safety risks may as easily be disproportionately distributed among males—for example, the chemicals and safety hazards that typify mining and heavy manufacturing industries—or among ethnic or racial minorities that, through custom, tradition, or social prejudice, are disproportionately represented in certain workforces.

ORGANIZING FOR THE ASSESSMENT OF IMPACTS ON PUBLIC HEALTH

The assessment team is highly likely not to include personnel who are professionally trained in public health disciplines. In fact, expecting that the typical assessment team will include individuals trained in many of the disciplines that must be considered in a comprehensive impact assessment is quite unreasonable. Therefore, how can the assessment team identify the public health implications of all project impacts? How can the assessment team identify the "positive" impacts on the physical and social environment that can result in "negative" impacts on public health?

As in so much of environmental impact assessment, the answers to these questions lie not as much in the disciplinary expertise of assessment team members as in organizational skills that can facilitate the team's access to and utilization of knowledge and expertise that are available from sources external to the formal assessment team.

For the purposes of the following discussion, we assume that the responsibility for assessing public health impacts is delegated to a specific individual or subgroup of the total assessment team. The following guidelines are offered for integrating the efforts of those responsible for assessing impacts on public health with the efforts of those responsible for assessing other types of impact, and for conducting an assessment of public health impacts.

Procedural Guidelines

1. In the early phases of project development (i.e., predesign and locational phases), individuals responsible for assessing public health impacts should establish liaison with governmental agencies and organizations (including academic institutions) that have jurisdictional interest in public health issues and environmental quality. Establishing liaisons with groups who have direct access to health data and information relevant to populations residing in the proposed project area, as well as with groups who have previous experience with the type of project under consideration, is particularly important. The objectives of such liaisons are to identify likely types of public health impacts and project-relevant factors that might influence those impacts on the local population, and to establish access to practicing professionals who can assist in the assessment of impacts on public health.

2. Management of the assessment effort should insure that individuals responsible for assessing impacts on public health are immediately and fully aware of all impacts on the physical and social environment that are identified in the course of the assessment process, as well as all proposed mitigation measures. This immediate communication may be accomplished by in-house memos, written reports, or oral progress reports. Mechanisms should also be established by which personnel responsible for assessing impacts on public health may initiate requests for information from other members of the assessment team.

3. Individuals who are responsible for assessing impacts on public health should continually update other members of the total assessment team on public health and safety issues related to project development. Information transmitted to the total assessment team should include (a) sources of risks and (b) environmental factors (physical and social) that can influence specific risks. The objective of this approach is to sensitize other members of the assessment team to the roles that specific environmental components (e.g., dusts, runoff, noise) and dynamics (e.g., biomagnification, air dispersion of particulates) can play in human health. This sensitization will increase the probability that other team members will keep individuals responsible for public health assessment informed of key findings that might be relevant to the health assessment.

4. All phases of project development (from early design through operation and maintenance phases) should be considered for their direct and indirect influence on public health. This long-term assessment requires an evaluation of (a) each activity undertaken in each phase, (b) the possible direct and indirect impacts of each activity on environmental components and dynamics that may play causal or mediative roles in public health

impacts, (c) the acute and chronic manifestations of health and safety impacts, and (d) the social equity of those impacts.

5. All environmental standards for public health (e.g., water quality standards, air quality standards, noise standards) should be used by the assessment team in evaluating potential health and safety impacts. Environmental parameters (including physical and social parameters) that are likely to be affected by project development, for which no safety or health standards have been promulgated, should be evaluated in conjunction with recognized health authorities and in light of the developing scientific literature.

In following the general guidelines just presented, assessment personnel will find their task greatly facilitated if they organize their analytical and integrative efforts to answer the following questions. Note that the form of these questions is the basic form of the historical reportorial questions designed to enhance the precision, conciseness, and relevance of information-gathering and writing tasks.

1. What are the specific physical and social consequences of each activity in each phase of project development (e.g., an increase in turbidity of water during construction, atmospheric dispersion of hydrocarbons during the operational phase, introduction of pesticides into terrestrial habitats during maintenance operations)?

2. Where will these consequences and their subsequent (i.e., indirect) effects be manifested (e.g., in surface supplies used for potable water, in agricultural soils, in higher order consumers in wetland food webs)?

3. How will these consequences be specifically manifested (e.g., increased concentration of heavy metals in surface soils, increased rates of morbidity of carnivorous lacustrine fish species, increased concentration of pathogenic bacteria in water table wells)?

4. What qualitative or quantitative factors can influence the manifestation of these consequences (e.g., climatic factors such as prolonged drought and elevated daytime temperature, population density of higher order consumers)?

5. Which human populations are likely to be subjected to these consequences of project development (e.g., children playing on grounds immediately downstream from atmospheric emissions, families dependent on hunting and fishing, consumers who buy specified agricultural products)?

6. How can the consequences of project development "enter into" each human population (e.g., absorption through skin, respiration, sensorial inputs, ingestion)?

7. What social, physiological, psychological, or other factors can influence the effects of these inputs into each human population (e.g., age, sex, occupation, behavior, pre-existing health conditions, diet)?

8. Considering these factors, what types of effects are these environmental factors likely to have in each of the populations considered (lethality, morbidity, reproductive effects, genetic effects, behavioral effects, learning disabilities, developmental anomalies)?

CHAPTER 14

ECONOMIC IMPACTS

Various categories of economic factors have been used for purposes of assessing economic impacts, including local and regional factors such as:

- employment/shopping facilities
- residential property values
- property tax base
- displaced residents
- displaced businesses
- remaining businesses
- new businesses
- multiple uses of local resources

In some instances, priority has been given to those factors considered most pertinent to economic *structure* (e.g., employment by industry, public versus private sector income, economic base, income/wealth distribution). Sometimes priority has been given to factors considered most pertinent to economic *conditions* (e.g., income, employment, wealth).

Although such factors, and categories of economic factors, are important in any consideration of the economics of a particular area, they do not describe the economic dynamics of that area. To describe impacts on an economic system, we must understand the dynamic interrelationships among individual factors and groups of factors.

The economic literature is replete with diverse definitions of "economic system." Nonetheless, some authors have considered that the differences merely reflect different degrees of emphasis on commonly recognized aspects of economic systems. For example, some researchers have suggested that one general definition of "economic system" describes it as an arrangement of people and organizations that functions to provide the goods and services required by the public. In this definition, preeminence is given to the economic system as a **means of meeting the needs of the public.** Another definition describes the economic system as an arrangement of people and organizations that functions to meet the needs of the public or the needs of its own organizations. In this definition, emphasis is given to the **self-serving character of key economic powers,** such as large corporations.

With respect to the broad environmental context of economic systems, the economic system is sometimes viewed primarily as a means by which irreplaceable resources are transformed into transient monetary wealth. Sometimes, this system is considered not so much the means but the actual determinant of the transformation of raw resources into goods and services. For the purposes of an overview of economic impact assessment, highlighting various attributes and functions of economic systems, as addressed by a wide diversity of definitions, is useful.

- An economic system is composed of **interacting individuals and organizations.**
- An economic system **regulates the transformation of raw resources** into socially required goods, services, and monies.
- An economic system **accomplishes the social distribution** of goods, services, and monies.
- An economic system **influences social perceptions** of what goods, services, and monies are required.

Although we can speak of the economic system of a nation as a whole, more relevant to the objectives of impact assessment is identifying the particular form of the general national economic system that is manifest in the local project and surrounding regional area. This form will vary with (1) the nature of individual and organization interactions; (2) the nature of resource transformation into goods, services, and monies; (3) the local pattern of distribution of goods, services, and monies; and (4) local perceptions of required goods, services, and monies.

Regional variations in these economic attributes and functions can be immense, but certain dimensions can be identified that can be evaluated on a site-specific basis. Some of these dimensions are identified in Table 14.1. Note that the dimensions identified in this table are only examples. Different assessment teams may find it more convenient to use somewhat different categories to reflect the real economic conditions and structure of a particular locality more precisely.

The importance of assigning real dimensions to general attributes and functions of local economic systems canot be overstressed. First, such dimensions are the basis for identifying types of data that should be collected by the assessment team. Second, these dimensions are basic tools for evaluating and relating specific types of data to system dynamics. For example, one dimension of local individual and organizational interactions identified in Table 14.1 is *dependency relationships*. Dependency relationships may in-

TABLE 14.1   Examples of Economic Attributes, Functions, and Dimensions

General attributes and functions of economic systems	Dimensions of attributes and functions of economic systems
Nature of individual and organizational interactions	Dependency relationships among individuals and organizations; competitive relationships among individuals and organizations; long- and short-term duration of relationships; alternatives to ongoing relationships; mobility of participants; current and projected trends
Resource transformation into goods, services, and monies	Nature of local resources currently utilized; external sources of resources; availability of workforce; availability of untapped resources; perceived desirability of current and potential resource utilization; dependability and adequacy of external resources; transportation requirements
Distribution of goods, services, and monies	Patterns by sex, age, race, educational background, ethnicity, socio-economic status; population projections; community services; per capita and family income; hiring practices; labor and capital costs; profits
Perceptions of required goods, services, and monies	Demographic differences; source of perceptions; reinforcement of perceptions; community values and life-styles

clude (1) family member dependency on head(s)-of-household, (2) dependency of small markets on the economic stability of neighborhood households, (3) dependency of local manufacturers on regional and/or extraregional markets, (4) dependency of local public schools on local property taxes, and (5) dependency of local and regional populations of young people on local businesses for jobs.

Types of data and information that are important for understanding the nature of such relationships in local areas may include (1) annual family and *per capita* income, family size, sex, age, and other characteristics of local heads-of-households; (2) number and distribution of household units, current and projected household debt, small-business bankruptcy, and new business ventures; (3) adequacy and costs of transportation facilities, local tax incentives to business, and regional competition; (4) school enrollment, emigration and birth rates, tax base, and cost of living; and (5) educational and training facilities, alternative local and regional employment, current and projected labor supply and demand, and wage levels. Once data and information required by the appropriate dimension of economic systems have been collected, the assessment team must seek to understand how all dimensions of the local economic system may be interrelated, that is, they must establish plausible causal chains.

Although no universally accepted, objective formula is available for integrating specific data and information on economic variables into a precise systems diagram of an economic system, a systems overview of a local economic system (Figure 14.1) can be achieved if the assessment team adheres to a few basic precepts:

- The nature of individual and organizational interrelationships in any local area does influence the type of resource transformations into wealth within that area.
- The type of resource transformations does influence the patterns of distribution of that wealth.
- Social perceptions of required wealth are influenced by social interrelationships, the types of resource transformations, and the patterns of wealth distribution.
- Social perceptions of required wealth may act as feedback modulators to each of these economic attributes and functions.
- National and regional economic factors also influence the attributes and functions of local economic systems.

With respect to the practical usefulness of this systems approach to economic impact assessment, note that the problem of identifying and describing the basic elements of economic systems on a site-specific basis is not a new problem invented by the regulatory need for environmental impact assessment. This problem has been faced and dealt with for many years, especially in the field of economic anthropology in which—despite substantial disputes concerning the proper subject matter of economics, the nature of economic systems, and how economic systems should be studied—wide consensus exists that **economic activity is a social process and cannot be simply circumscribed by the type of technology employed within a specific society.** In this regard, impacts on jobs, on income, or on any other single economic attribute are not themselves impacts on economic systems. As depicted in the system overview presented in Figure 14.1, the assessment

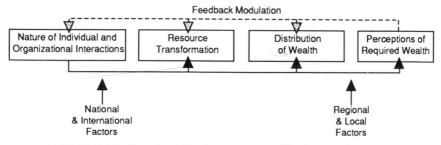

FIGURE 14.1 Overview of basic components of local economic system.

team should make an effort to evaluate individual economic impacts in light of ongoing dynamic interrelationships among social and physical components and dynamics of the total human environment.

For example, a particular project may result in short- and long-term increases in the number of jobs in a project area. If the increase in jobs is presented only as an individual impact, this isolated piece of information is unlikely to help improve the decision-making process in project planning and development. Only when the potential for an increase in the number of jobs is related to other social and physical components and dynamics in the total human environment can the impact be evaluated for purposes of decision making. For instance, what will be the consequence of an increased number of jobs for local business, for land development in the area, for recreational demand, and for population mobility? Who will get the higher paying and the lower paying jobs? What are the other monetary costs that might be associated with these specific jobs, for example, costs related to medical care as a result of occupational disease and disability?

Answers to such questions help decision makers determine the real significance (or insignificance) of a project-mediated increase in the number of jobs. Of course, such questions cannot be asked, nor can answers be given, if the assessment team fails to take a systems overview of categorical economic impacts.

ENVIRONMENTAL BENEFITS AND COSTS

Impact assessments typically reduce the complexity of economic impacts to a relatively simple model of project costs and project benefits. Therefore, the assessment of economic impacts tends to become an exercise in cost–benefit analysis, in which various project alternatives (including the no-action or no-build alternative) are evaluated for their potential to achieve a maximum of benefits at minimal costs. Because everyone has at least an intuitive appreciation for what constitutes a benefit and what constitutes a cost, a cost–benefit approach to impact assessment appears to be logical. However, like the dichotomous mode of thinking that results in the categories of "right" and "wrong," the categorical use of project costs and project benefits can be greatly misleading in a world composed more often of shades of gray than of black and white.

The basic problem is that cost–benefit analysis typically employs meanings of "cost" and "benefit" that are so narrow that they often become irrelevant to the real community to which they are addressed. For example, for purposes of a cost–benefit comparison of project alternatives, benefits are often defined by how much people would be willing to pay for project outputs. Which people are included in the survey? Should the willingness to pay a certain sum of dollars be extrapolated from a highly developed urban database or from a rural database?

In evaluating benefits, primary attention is often given to such project effects as decreases in the amount or costs of resources required to produce goods and services. These *allocative* benefits are statements of sum totals of dollars saved, without reference to whose dollars they are. *Distributive benefits* or *costs,* which specify precisely who receives which benefit and who pays which cost, are typically omitted. Also, costs associated with low-probability but disastrous events (e.g., dam failures) are typically omitted in the usual type of cost–benefit analysis. Finally, some project-mediated effects canot be easily assigned monetary values (i.e., *intangible costs and benefits*), for example, "aesthetic deprivation." Although monetary values have sometimes been assigned to such intangibles, much concern has been raised over the seemingly arbitrary nature of such valuations.

Cost–benefit analyses that focus on allocative benefits of proposed projects, that also ignore distributional effects and also arbitrarily assign dollar values to the intangibles of daily life, have been characterized as having very limited utility for selecting among alternative courses of action. Of course, the limited utility of cost–benefit analysis is also directly proportional to the failure to consider *hidden costs,* such as increased insurance rates that might occur as a result of suburban development, medical costs of occupational disease associated with newly developed industries and services, and costs of social services (e.g., potable water supplies, schools, fire and police protection) associated with infrastructure development.

In an effort to evaluate the comprehensiveness of a cost–benefit analysis, consider the following questions:

1. Do identified benefits and costs include benefits and costs for which market prices exist that actually reflect local social values and economic conditions?

2. Do identified benefits and costs include benefits and costs for which market prices exist that do not reflect local social values and economic conditions?

3. Do identified benefits and costs include benefits and costs for which no market prices exist, but for which appropriate social values can be approximated in monetary terms by inferring what consumers would be willing to pay for the product or service?

4. Do identified benefits and costs include benefits and costs for which imagining any kind of market process capable of registering a meaningful monetary valuation would be difficult?

5. Does the analysis identify the specific beneficiaries (present and future, local and remote, majority and minority) of economic benefits?

6. Does the analysis specifically identify those populations (present and future, local and remote, majority and minority) that will pay the economic (and other) costs?

7. Does the analysis include a rationale for evaluating the desirability or undesirability of the projected redistribution of economic resources in terms of local social values?

SOME EXAMPLES

Most often, responsibility for the assessment of economic impacts has been given to a highly specialized subgroup of the overall assessment team; interactions among this and other subgroups with other assessment responsibilities have been largely lacking. Superficially, this condition appears to be a rational approach. After all, wildlife biologists are not generally known for their insatiable curiosity about the difference between allocative and distributional aspects of economic systems, nor are property assessors known for their expertise in aquatic or terrestrial ecology. However, two facts indicate that the assessment of economic impacts solely by individuals with "economic expertise" is unwise.

First, impact assessment calls for interdisciplinary assessment. Second, a truly comprehensive evaluation of project-related costs and benefits includes consideration of intangible as well as tangible effects and distributional as well as allocative effects of project development, neither of which can be surmised by individual experts working in isolation from other members of the assessment team.

The following precepts underscore the importance of integrating the assessment of economic impacts with the assessment of impacts on the physical and social components of the total human environment.

- Project impacts on the physical and social environment can result in changes in the short- and long-term economic conditions and structure of local and regional areas.
- Project impacts on the economic conditions and structure of local and regional areas can result in changes in the physical and social environment.

Example 14.1 Secondary development that occurs as the result of the construction of a new highway may result in the progressive conversion of local wilderness areas to developed land. If existing wilderness areas have a high carrying capacity for game animals, a high potential for the production of timber for a lumber industry, or an existing or potential value as wilderness camping areas, then subdivision development can result in future economic constraints. Of course, immediate and long-term economic gains as a result of subdivision development (i.e., in terms of local tax base and consumer spending) may be much greater than the monetary return from alternative uses.

Example 14.2   Specific health-care costs can be attributed to various aspects of project-mediated increases in the industrial base of a community. Some of these costs are associated with workplace safety and exposure to hazardous chemicals and situations. Some are associated with project-related releases of hazardous chemicals to the environment, including atmospheric emissions, contamination of ground- and surface water, and accidental spills, with subsequent exposure of domiciles, persons, and food and water supplies.

Example 14.3   Long-term costs of project-mediated subdivision development might reasonably be expected with respect to growing needs for potable water, waste disposal, community services, and recreation. Other long-term costs might be associated with the social disorganization of existing life-styles and the consequent development of youth gangs and criminal subcultures.

Whereas the preceding examples focus on some economic impacts to be considered as a result of project-mediated changes in the physical environment, considering the converse, that is, how project-mediated changes in economic conditions can cause changes in the physical and social environment is also important.

Example 14.4   Subdivision development that may result from project development in one area can create consumer demands that will influence economic development in another area. Such demands might include demands for recreational facilities, manufactured goods, and services. The meeting of these demands can result in (1) reduction in wilderness area (e.g., through the construction of ski runs, boating and bathing areas) and (2) increase in pollutional loading of aquatic and terrestrial ecosystems (e.g., through industrial processes, waste deposition, transportation).

Example 14.5   Land development projects as well as national and regional improvements in bulk transportation and farm technology may result in a dramatic local decline in the acreage previously used for marginal agriculture and husbandry. Declines in a local dairy industry, for example, will typically result in a decrease in the number of acres set aside for haying. With the removal of haying activity, the land will revert to its ecological successional pattern, with short- and long-term changes in vegetation and wildlife. In some instances, however, the economic loss to the farmer caused by a local market decline in hay, coupled with increasing land taxes, might result in local farmers selling topsoil, with consequent long-term impact on local terrestrial ecology as well as on surface and groundwater aquifers.

The preceding examples by no means exhaust the possibilities of direct and indirect project impacts on the economic conditions and structure of local and regional areas. However, they are sufficient to demonstrate an elementary fact about the economic impacts of projects, namely, that in the broadest sense of economics, economic impacts of project development cannot be separated from impacts on the social and physical components and dynamics of the total human environment. Thus, the assessment of economic impacts must be conducted in parallel with and integral to the assessment of social and physical impacts of project development.

Example 14.6   A project-mediated increase in available jobs, especially low level jobs requiring little if any skills or experience (fast food services, sales counter services, janitorial services, etc.) are most often considered beneficial to local youth. However, in many instances high schoolers spend increasing time at the workplace and decreasing time in the school room, with subsequent serious detriment to their student status. In such instances, the supposed economic benefit of a proposed project might reasonably be considered to be an actual long-term economic cost in terms of the student's aborted schooling and subsequent relinquishment of a future more productive and better remunerated employment.

Example 14.7   The acquisition of property, either through sale or condemnation, might result in the displacement of families. In some jurisdictions, specific legal requirements exist that families displaced through condemnation proceedings must be provided with comparable housing that, in addition, must meet specific standards regarding safety and sanitation. Thus, under some circumstances, a family might realize a significant improvement in housing as a result of the taking of its original household. However, the newly provided, improved household may represent an economic asset previously unavailable to the family's creditors and, in the absence of additional specific legal protection, may become attached with a lien by those creditors.

Example 14.8   As discussed in Chapters 7 and 9, a variety of project-related activities (e.g., excavation, land clearing) can result in short- and long-term impacts on downstream or offshore fisheries. Economic costs associated with such impacts may be significant, especially for estuaries, which are often important nurseries for offshore commercial fisheries.

Example 14.9   Estuaries and other shoreline wetlands may also play a critical role in the biological treatment of wastes contained in land runoff. The filling of such wetlands and subsequent loss of this important function may also be considered an economic cost of project development that can be directly determined on the basis of the cost of comparable treatment carried out in a water treatment plant.

Example 14.10   Land development projects in lacustrine watersheds are highly likely to exert a long-term influence on the eutrophication of surface waters (Chapter 7). Depending on the use of lacustrine resources, enhanced eutrophication may result in significant costs, that might include decreases in shoreline land value, due to the loss of water quality and the emergence of nuisance vegetation. Costs might also be incurred because of chemical, biological, or other treatment techniques employed to sustain water recreation or to meet regulatory requirements.

SOME GUIDELINES

Although a variety of techniques is available for assigning monetary value to the types of impacts considered in the preceding examples, the fact remains that so-called "intangible costs," which are of real interest and concern to real people, cannot be easily assigned monetary value. Thus, real costs with respect to such issues as privacy and aesthetics and such emotions as frustration, anger, helplessness, and confusion are typically ignored in any cost–benefit analysis of project development.

In light of the methodological difficulty of dealing with intangible costs, we suggest that these intangible costs be specifically addressed in the context of social impacts, that is, as nonquantifiable impacts on personal and interpersonal components and dynamics. When such personal and interpersonal impacts can be directly linked to monetary costs (e.g., health-care costs, law enforcement costs), they can also be specifically addressed in a formal cost–benefit analysis. This approach underscores **the importance of viewing a cost–benefit analysis as an adjunct to rather than a substitute for the assessment of social impacts.**

Additional guidelines that should be considered by assessment teams in pursuance of a comprehensive assessment of economic impacts include the following:

1. All direct and indirect impacts on physical and social components and dynamics should be evaluated for their economic consequences.

2. All direct and indirect impacts on the economic conditions and structure of local and regional areas should be evaluated for their consequences on the physical and social components and dynamics of the local and regional environment.

3. Comprehensive economic analyses of project alternatives (including the no-action or no-build alternative) should specify and justify the valuation of the short- and long-term allocative and distributional effects of project development.

4. All economic effects of project development should be evaluated

in light of the goals and objectives of pertinent legislation and regulations, as well as local and regional values.

5. Economic analyses should be inclusive of all phases of project development, from the earliest planning phase through operation and maintenance phases.

6. All assumptions and limitations of economic analyses should be clearly identified and discussed with respect to pertinent legislation, available data and information, local and regional values and plans, and project objectives.

7. No one economic criterion (e.g., an increase in jobs) should be explicitly or implicitly offered as the single most important criterion of the desirability of the proposed project.

CHAPTER 15

CULTURAL IMPACTS

Historically, the term *culture* evolved out of studies of small, essentially self-sufficient social groups (most often called *tribes*) that typically lived in relative isolation from other groups. Although many specific definitions of culture have since come into use, each one reflecting the particular interests and emphasis of individual investigators, the term is generally used to refer to the distinctive, socially shared constellations of rules governing the behavior of a people, their beliefs, their values, and even their emotions.

In applying the concept of culture to densely populated, geographically extensive, pluralistic societies, numerous conceptual problems arise if only because people who live within these societies often demonstrate such a wide range of diversity in behavior, values, and beliefs that imagining any significant commonality that is imparted by a mutually shared enculturation process is difficult.

The recognition of blatant diversity within the supposedly same culture is reflected in the term *subculture,* which implies that a smaller group of persons shares some of the learned behavior of a more inclusive society but differs significantly from that larger society in specific ways. Therefore, at different times, we refer to a *criminal subculture,* to a *drug subculture,* to occupationally based groupings such as the *military subculture,* and to socioeconomic, religious, ethnic, and racially based subcultures.

For purposes of impact assessment, the "correctness" of any particular definition of culture need not be determined. Just as physical scientists have continually devised new meanings of such concepts as "niche," for which dozens of specific definitions and different uses are common, so have social scientists continued to devise new meanings for the term "culture." However, we can also expect that such new meanings will still be built on a broad consensus on the following critical aspects of social life addressed by the culture concept:

- the social sharing of behavior and meanings,
- the critical role that social institutions play in transmitting and enforcing cultural rules from one generation to another, and
- the continual adaptation of human behavior and meanings to new environmental conditions.

These aspects of social life differentiate between cultural impacts and social impacts.

Impacts on a person's or group's behavior are social impacts; impacts on the socially transmitted rules that teach people how to behave, what is proper and improper behavior, are cultural impacts. Impacts on a family with respect to economic well-being are social impacts. Impacts on the role of the family as an institutional means of instilling particular values and beliefs are cultural impacts. Impacts that might affect the manner in which available natural resources ares utilized are social impacts. Impacts that result in social adaptations to totally new and different resources are cultural impacts.

In short, social impacts are impacts on people; **cultural impacts are impacts on the social means by which humans learn to become people.** Therefore, contemporary phenomena such as the *information superhighway,* the *sexual revolution,* and the movement for *women's rights* are often referred to as not simply social phenomena, but culturally significant developments that, like past historical developments of urbanization and subsequent suburbanization and so many others, alter the way in which humans see and understand themselves and their surroundings.

CROSS-CULTURAL CONTEXT OF IMPACTS

Many pluralistic societies contain within their geopolitical jurisdiction aboriginal peoples (or native peoples) who are the original residents of a land. Often, such collateral cultures are utterly distinct from the various cultures that constitute the governing pluralism. In many instances, collateral cultures are socially or geographically isolated; in some cases, little if any barrier exists to interactions between collateral and governing cultures.

Projects undertaken to meet the needs of a governing society often result in devastating consequences for an isolated collateral society, most often because of factors such as the extensive removal of resources on which the collateral society depends, the introduction of diseases into an aboriginal population with no natural immunity to those diseases, their sudden access to new technologies and alternative life-styles, and the outright mayhem and murder that a technologically more advanced society all too frequently perpetrates on the less technologically developed.

Even in the absence of a collateral culture, project development may result in impacts on the diverse cultural components of a pluralistic society, with possibly profound ramifications regarding the relative distribution of those impacts among majority and minority cultures.

Impacts on Collateral Cultures

Example 15.1   On the basis of a 20-year plan for national development, extensive areas of a virgin rain forest are opened for commercial lumbering as well as for the development of a highway system to service new mining operations. Both lumbering and mining operations occur in an area populated by a number of separate aboriginal groups that share the same culture and earn their sustenance by a combination of hunting (small mammals) and food gathering.

Often overlooked by people who, by virtue of their own society's high technology, obtain their food from a technology-intensive fossil fuel-subsidized agriculture is the definite limit to the number of persons that can be supported by each square mile of forest. In other words, the forest has a certain carrying capacity (Chapter 8) for humans because it has a certain carrying capacity for the types of animals and plants on which humans depend. A reduction in the extent of the forest is therefore equivalent to a reduction in the number of humans who can earn their sustenance by hunting and food gathering.

When a reduction in the carrying capacity of a resource for a particular wildlife species is evaluated by wildlife biologists, the approach is to calculate the reductions in population density that will occur through starvation or attendant illnesses. The approach is *not* to consider that the stressed population will move to new habitat, because that new habitat is likely to be already utilized by other populations. Similarly, reducing the carrying capacity of a forest for humans must translate into morbidity and mortality rates for that population.

Although increases in morbidity and mortality are potential impacts on the aboriginal population, they are not cultural impacts. Of course, carried to the extreme, the death of all members of the collateral culture is equivalent to the loss of that culture. However, even when whole populations are not lost, increased rates of morbidity and mortality can lead to long-term and possibly irreversible changes in culture. After all, when faced with starvation, humans typically do not willingly lie down and die. If hunting and food gathering suddenly proved to be insufficient means for obtaining food, other means would be explored. Certainly the opening of the forest, which is the potential cause of their eventual starvation, would also be a potential source of alternative ways to obtain food. For example, hunters and food gatherers might become "performers" for tourists and anthropologists whose money can be exchanged for canned goods produced by the governing society. Perhaps these aboriginal peoples could become guides to construction crews, laborers in new mines, or launderers in construction camps.

Ultimately, however, these indigenous people are only likely to survive by abandoning the old ways of life and adopting new ways. In such a manner, in even one generation, hunting and food-gathering disappear as do the values learned in their practice and the kinship-enforced rules of sharing the kill, the rituals performed for the hunt, the status of elders who are the teachers of the hunters and the gatherers, and the distinction between male and female roles that underlies much of their daily behavior. These changes are the potential cultural aspects—irreversible changes in the way a people perceives and experiences itself and the world. The anthropologist calls these changes acculturation: the dissolution and eventual death of a culture.

A project need not remove essential resources to have dramatic impacts on the affected culture.

Example 15.2 A national policy is enunciated regarding the need to decrease reliance on imported energy sources by increasing the development of internal resources. However, a large portion of the governing nation's reserves of oil, coal, and uranium are owned by diverse collateral societies, which hold deed on behalf of their respective aboriginal nations. Thus, the policy promoting energy self-reliance not only results in the increase in private investment in the licensed development of the governing state's public lands, but also increases the market pressure on collateral societies to develop their own collective resources.

Historically, the collateral nations have no concept of individual ownership of land or other natural resources within their cultural experience, whereas this concept is central within the cultural experience of the governing nation. Without historical precedent to serve as a determinative reference, discussions within the councils of the collateral nations devolve into serious conflict, which is quickly polarized to those favoring a traditional view of land, which is inimical to any concept of land as something to be bartered for personal gain, and the contrary view, which is consonant with the values of the governing nation. The disagreement broils over to open armed conflict and lives are forfeit.

Whatever the final resolution of such a conflict might be, the impact of the governing society's policy on the collateral culture is clear: an abrupt escalation of a long-standing acculturation in the collateral society, which is engendered by that society's physical proximity to a governing society with significantly disparate values that cannot be avoided or ignored. Once any culture evolves into a condition of severe internal conflict regarding central cultural values, that culture has already evolved into a new culture; the old one, regardless of the artifacts of its former self that might be preserved as heirlooms, is dead.

Example 15.3 A governing nation implements extensive mining operations that result in a metals-contaminated leachate that, through a water table aquifer, enters a stream and is subsequently released to an estuary far removed from the mining operations. This estuary is an important nursery to offshore fisheries that are used by an island-based collateral society as a major food source. However, the detritus-based food web of the estuary results in the trophic magnification of toxic metals in those fisheries, thereby effectively removing this food supply for the foreseeable future.

The collateral society is composed of shoreline and inland communities that, through a series of kinship-based and religious institutions, exchange fish and agricultural produce, thus insuring a balanced diet in both shoreline and inland populations. The effective removal of the offshore fisheries results not only in a serious imbalance in diet, but also in the dissolution of marriages and other kinship mechanisms that are integral not only to the exchange of fish and agricultural products but also to core cultural values that have traditionally insured stable and harmonious interactions throughout the island community. The acculturation process begins abruptly, punctuated by escalating family discord, institutional collapse, and subsequent internecine warfare.

Although these examples illustrate some of the mechanisms by which collateral cultures may be significantly altered by the actions of a governing society, they do not begin to describe the extent of the historical record of utter decimation and degradation consciously and unconsciously visited on aboriginal cultures by intruders. At a time when many individuals would extol the effort to protect threatened and endangered wildlife species, the protection of threatened and endangered human cultures is certainly of no less import or urgency.

However, even the most concerted efforts to protect a culture cannot guarantee against a continuing cultural adaptation to essentially uncontrollable environmental changes. Cultures, after all, do evolve. A good example of ongoing cultural evolution that affects many contemporary societies is the relationship between traditional kinship systems and the requirements of global economics. Of particular importance in social dynamics is the role of kinship, the manner in which individuals identify themselves and each other with respect to their relatives. As indicated in the preceding examples, the rules of kinship in nonindustrial aboriginal cultures are typically the basis of an individual's access to and use of those essential social interactions that define power, obligations, duty, and authority, and thereby continually modulate socially accepted goals and socially accepted means for achieving those goals.

The kinship systems of nonindustrialized aboriginal cultures typically give precedence to the parent–child relationship rather than the parent–parent relationship, that is, to the consanguineous (or genetic) descent of a person rather than to the person's affinal (or marriage) association. In such cultures, descent is also typically traced in a unilineal fashion, along the female (matrilineal) or the male (patrilineal) line of descent, as in a *clan*. Unilineal kinship systems identify the individual as belonging to a relatively large number of socially selected genetic relatives that, extending over limitless generations, becomes the basic unit of social organization. This collectivity, and not the individual, exercises ownership of property and determines the distribution of goods.

Bilineal kinship systems, which identify the child within the context of a nuclear family and give equal importance (or disregard) to the families of both sides of the marriage, are characteristic of much of the Euro-cultural tradition. In such systems, the basic unit of social organization is the numerically small, temporally limited, nuclear family.

Because the historical development of industrialization is so intertwined with the societal evolution of Europe and the United States, separating essential aspects of modern industrial life from the exigencies of the nuclear family is difficult. As societies that have a unilineal kinship tradition increasingly choose to evolve into industrial societies, the tendency is to adapt to a bilineal mode of kinship—a process that, in comparison to the insulated comfort of a populous and multigenerational descent-based family, increases the exposure of the individual to a plethora of social pressures and influences that he or she cannot easily understand or even begin to control. Personal confusion, frustration, anger, fear, and even alienation may therefore be considered among the prices paid by many individuals for their determined entrance into an increasingly integrated global economy.

Impacts on Constituent Cultures

National political identity, which is itself a relatively recent phenomenon in world history, often obscures the fact that a nation is most often composed of diverse constituent cultures that, despite broad social discourse and interaction and even the periodic crystallization of a perceived common cause, manage to maintain much of their original content over very long periods of time. Although this statement does not suggest that these cultures do not change, certainly the evolution of socially interactive cultures is not so much an innately driven rush toward assimilation as it is, perhaps, a stochastic meander between confrontation and accommodation.

The sense of pluralism always requires cultural differences within a populous and geographically extensive population. Some of these differences reflect the original contents of diverse but constituent cultures; some, the social evolution of those particular cultures; and some, those new cultural patterns purposely or unconsciously fashioned to address the new realities that new generations must face. Although much good reason exists to focus assessment efforts on those constituent cultures of a society that are historically defined with respect to ethnicity and race, cultures defined with respect to attributes that are not subsumed or otherwise limited by ethnic or racial considerations must also be considered. After all, social science has long addressed certain groups as having distinct cultures, including street gangs and incarcerated prisoners. Often spoken of as subcultures, such cultures are probably best described as "situational," that is, behavioral rules, values, and beliefs provide the basis for coping with a persistent condition that is largely influenced by social factors.

Example 15.4 A strip development project is undertaken along a major highway that runs through a rural township located between two distant metropolitan areas. The project includes a motel as well as several retail clothing stores and fast-food establishments. For 9 months each year, the total population is primarily composed of students who attend a local college. The full-time residents of the town are almost entirely of the same ethnic minority, whereas the vast majority of the college population (including students and faculty) is of a different ethnic majority.

On completion of the project, the employment patterns associated with the new development are clearly highly correlated with ethnicity. The lowest paying jobs, which also require the least interface with the general public (e.g., hotel maids, dishwashers, and low-skill maintenance jobs), are held entirely by the ethnic minority.

In such a situation, mutual stereotypes are highly effective means of maintaining social and cultural distance. Such stereotypes, which are typically highly pejorative, are particularly effective barriers when they are essentially self-fulfilling prophesies. For example, a minority that does not speak the language of the majority may be stereotyped as "too mentally lazy to learn anything new." Hiring practices that effectively isolate that minority in situations that do not require new language skills therefore help fulfill the prophecy of mental laziness. Those same practices may also help fulfill the minority's stereotype of the majority as "thinking themselves too good to do honest work."

In addition to social impacts (e.g., outright ethnic confrontaton and conflict), such ethnically biased hiring practices may also result in significant cultural impacts, such as:

- a strengthening of the cultural bias of both minority and majority cultures,
- an enhanced proliferation of the majority's negative bias against the minority (e.g., as in the "infection" of majority students who have had no previous contact with the minority), and
- the potential gradual incorporation of negative self-esteem in the minority's cultural self-image (such as feelings of an inherent inadequacy).

Similar types of impacts may be realized when the distribution of jobs is directly correlated with situational cultures.

Example 15.5   An industrial development includes an electrocoating operation that involves highly repetitive work with small metal parts, including degreasing, spray coating, rinsing, and drying. The workforce is derived from a local population of youthful unskilled labor that is willing to work under relatively dirty conditions for minimal wages. Overtime work is commonly required.

In this case, the workforce may include persons of any number of different races and ethnic backgrounds (including majority and minority). The unifying attribute of this situational culture is *low education—low skill attainment,* a condition increasingly recognized in the United States as defining a "permanent underclass" of citizens that has been essentially left behind by the rapid developments of high technology and global economics. The salient feature of such an underclass is the almost impenetrable social and economic isolation of workers who, by the nature of their employment, do not learn new skills while on the job and cannot, because of the absolute constraints of limited time, money, and experience, seek out nonremunerative training in new, more marketable and lucrative skills.

As in the case of persistent racial and ethnic isolation, the social isolation of such a workforce in an essentially affluent pluralism is tantamount to the forced development of new cultural norms that are likely to be largely inconsistent with, if not directly contrary to, those that characterize historical cultures based on ethnicity and race. In this sense, the development of workplace facilities that benefit from such an underclass might realistically be considered yet another example of the exploitation of one culture for the benefit of another, that perpetuates cavalier assumptions concerning the expendability of certain human populations and the superiority of others.

Example 15.6   A major expansion of an airport is undertaken that requires the consolidation of a number of undeveloped plots of public land in the general vicinity of a densely populated ethnic community. This community represents the second or third generation of an original rural population that migrated to the city to take advantage of a rapidly growing job market for laborers.

Although the consolidated properties have long been considered undesirable for purposes of commercial or residential development, the surrounding community has continuously used various portions for growing vegetables. Different neighborhoods within the community cooperate in the tending of their own gardens. In fact, each of the seven communal gardens serves as a central gathering area where personal, familial, and neighborly interactions are often more important than the actual agricultural effort, especially with regard to the resolution of festering conflicts that periodically threaten their normal tranquillity. These social interactions are generally restricted within each garden group, but a strong tradition of friendly agricultural competition exists among them that is manifest in periodic celebratory visitations of each others' gardens. In the course of these visitations, cultural bonds between physically more distant residential neighbors are reinforced.

From the perspective of the airport development authority and city administration, the taking of these properties is a simple and highly practical matter that involves land parcels already owned by the city. Structures do not have to be demolished; people do not have to be relocated. From the perspective of the ethnic community, of course, these takings represent a devastating blow to its very way of life.

All cultures are means for achieving the physical and social survival of people. Except in extremely small societies, no one person knows all aspects of his or her own culture nor does everyone have a precisely equal knowledge of any single aspect of that culture. A culture is, therefore, inherently variable: not so much a collection of precise algorithms as a social toolchest of possible heuristics. This variability allows a culture to adapt to new conditions, that is, to evolve through a piecemeal process of acculturation by which certain cultural attributes are altered or even discarded and others are newly adopted or invented outright. Sometimes the changes are minuscule; sometimes they are dramatic. However, even during periods of pronounced evolution, some features of a culture persist that change more slowly than other features and, therefore, give some measure of continuity.

In the example just given, the ethnic culture is already in an obvious state of change, as reflected by the ongoing urbanization of originally rural life-styles. However, a core element of the original culture persists and is daily manifest in the use of communal farming as a means of resolving conflicts and strengthening social bonds. The airport project does not, therefore, cause or initiate the acculturation process; instead, it abruptly removes the primary cultural means of modulating the direction and rate of an in-progress adaptation to a new way of life.

NEED FOR A CULTURAL APPROACH TO IMPACT ASSESSMENT

Sufficient time and money are typically not available to an assessment team to undertake the type of detailed study conducted by a professional social anthropologist; neither are the time and money typically available for undertaking the type of detailed study conducted by a professional limnologist, or professional ecologist, or professional wildlife biologist. Budgetary constraints apply equally to all facets of impact assessment and, in themselves, cannot be used as an excuse for ignoring vitally important aspects of the total environment.

Rather than budgetary constraints, the more pertinent reason for the all too frequent failure to integrate broad cultural concerns into the assessment process is, in essence, a limited comprehension of social reality. All humans see and understand themselves and their environment in cultural terms. Culture constructs what we call reality, and culture informs us on how to deal with that reality.

If (as in Ex. 15.6) a vegetable garden is seen only as a piece of land used to obtain vegetables (which, after all, can be bought at a market) and not as a social life-support mechanism for a culturally evolving population, it is because one's vision is directed by one's own cultural experience, toward the reality of certain attributes of the environment, and away from the "unreal." Of course, we can understand another's perspective intellectually—see through his or her eyes what might not be seen otherwise. However, the new vision might be rejected for very rational reasons. After all, even if the garden meets a social need of a people, a different need will be met by the airport project. However rational, the reasons are also based on one's own cultural experience, which make compelling sense only among those who share similar cultural values.

The assessment process absolutely requires an empathetic understanding of alternative and even conflicting perspectives of the human environment. Without the studied practice of an empathetic perspective, and in the absence of any willingness at least to experiment with sympathy, the environment becomes simply that which the assessment team sees through the infinitely tiny window of its own limited but always distorting cultural experience.

CHAPTER 16

SPECIAL ISSUES: SOCIAL ENVIRONMENT

Special issues related to the assessment of impacts on the physical environment were discussed in Chapter 10. Those issues pertained primarily to special analytical techniques that are of increasing usefulness in the assessment process. In this chapter, the focus is more on managerial issues rather than analytical techniques because, historically, impact assessment has been conducted largely by personnel who have received their primary training in the engineering and natural science disciplines. Although this imbalance continues to be redressed, highlighting certain managerial aspects of the assessment process that are critical to the assessment of social impacts remains appropriate.

PERSONNEL

Although no accepted method exists for determining who may be best suited to undertaking assessments of social impacts, or how the assessment process is best conducted, some general questions should be addressed by the manager of the assessment team and by the public:

1. Is the disciplinary training of personnel actually representative of a cross-section of contemporary social, psychological, and behavioral disciplines?

2. Do personnel have actual field experience with social science techniques and methodologies?

3. Do in-house mechanisms exist to insure the communication of needs and findings among personnel responsible for assessing impacts on the social environment and those assessing impacts on the physical environment?

4. Are expenditures of time and money related to the assessment of impacts on the social environment comparable to those related to the assessment of impacts on the physical environment?

With respect to the disciplinary training and experience of personnel involved in the assessment of social impacts, we should highlight two concerns.

First, impact assessment requires skills (e.g., communication skills, analytical skills, integrative skills). Regrettably, although an individual may have a substantial academic background in a particular discipline, that individual does not necessarily also have the requisite skills for applying disciplinary knowledge to real problems and issues. Surely an academic background in social science disciplines that is composed primarily of introductory or "survey" courses **cannot be expected to have imparted** practical skills. Many undergraduate and even graduate programs in *environmental affairs* that purportedly attempt to cut across disciplinary barriers often seem to produce people who are more expert in reciting environmental aphorisms than in performing specific analyses and making informed recommendations.

Second, the most important reason for having personnel who are trained in the theory and application of social, psychological, and behavioral sciences is not that such personnel have the total expertise necessary for assessing social impacts but that they are likely to possess the attitudes, skills, and knowledge that will facilitate maximal professional use of sources of information that are external to the assessment team (e.g., governmental agencies, educational and research organizations, community and neighborhood groups). After all, nothing can be more contrary to the broad intent of the impact assessment process than an assessment team that becomes so impressed with its own expertise that it sees no need to consult with external organizations, groups, or individuals.

MANAGING THE ASSESSMENT EFFORT

Management practices that fail to insure intimate liaison between the assessment team and external sources of expertise and experience are poor management practices, regardless of the technical sophistication of the in-house team or whether the focus is assessment of physical impacts or assessment of social impacts.

How a comprehensive assessment of project impacts on the social environment can be conducted in the absence of specific information about project impacts on the physical environment is difficult to imagine, but this is often done because (1) when portions of the overall assessment are contracted out to consulting firms, coordinating interdisciplinary exchanges is difficult, and (2) even when the assessment is not contracted out to a consultant, management and technical personnel often fail to appreciate the interdependence of many social and physical impacts of project development. Technically or scientifically trained managers tend to enforce a rigid separation between what they consider natural science and "other" interests.

The responsibility for insuring meaningful interaction among personnel involved in assessing social and physical impacts is clearly the responsibility of the manager of the assessment project. Although the manager of an assessment project cannot absolutely insure the comprehensive assessment of social impacts, he or she can absolutely insure that comprehensive assessment will not be achieved. Too often, poor assessment is attributed to technical and scientific shortcomings of assessment personnel when, in fact, poor assessments are often the responsibility of a project manager who misuses the personnel at his or her disposal.

To provide an organizational framework for the comprehensive assessment of project impacts on the social environment, project managers should carefully consider the following general principles.

1. The scope of work for the assessment effort should be decided in joint consultation with assessment personnel responsible for assessing the social impacts as well as physical impacts of project development.

2. Allocations of personnel, budget, and other resources should reflect a balanced assessment of social and physical impacts and a major commitment to assessing interactions and interdependencies among social and physical components and dynamics.

3. Progress reports, review meetings, and other appropriate in-house mechanisms should be used for updating assessors of social impacts with information on physical impacts and *vice versa*.

4. Personnel who are responsible for assessing social impacts should clearly understand their responsibilities to coordinate their assessment effort with local and other organizations with jurisdictional or other interest in social components and dynamics.

5. No social impact should be ignored or otherwise considered superficially simply because it cannot be quantified objectively or be evaluated without consideration of political or other socially sensitive issues.

ANALYTICAL TASKS

Conducting analytical tasks efficiently is essential to conserving adequate time and budget for such important integrative tasks as the evaluation of the significance of project impacts and the design of appropriate mitigation measures. The efficiency with which analytical tasks are performed, especially with respect to the physical environment, is influenced by several factors including the familiarity of assessment personnel with promulgated environmental standards and guidelines. When such standards are plentiful and easily available, the process of impact assessment progresses very rapidly. However, with respect to most social components and dynamics, standards and guidelines simply do not exist.

Personnel responsible for assessing social impacts may, understandably, focus early in the assessment process on such potential impacts as the effect of noise on local residents and community functions, of project-mediated changes in air quality on public health and agricultural productivity, and of project-related activities on archaeological or historical resources. In each of these cases specific standards and guidelines can be used to identify and evaluate impacts. Nonetheless, by restricting analytical efforts on only these issues, the assessment team may discover that people in local and regional areas are primarily concerned with impacts on:

- community identity and life-style
- individual privacy
- population growth
- social disorganization, as manifest in criminal gangs and other deviant behavior
- personal social mobility
- sense of personal worth

No standards are promulgated by some regulatory agency that are relevant to these very real concerns, but this is not an excuse for the assessment team to ignore the potential relevance of project development to these concerns. These concerns are real, they do exist, and they do affect real people.

With respect to the quality of interdisciplinary interaction as a factor that can influence the efficiency of analytical tasks, we must recognize that no interdisciplinary assessment can take place in the absence of disciplinary knowledge and understanding. Interdisciplinary "sensitivity training" is not a substitute for multidisciplinary competence. Effective and efficient interdisciplinary communication and effort requires extensive multidisciplinary expertise and a commensurate level of oral and written communication skills.

Although expecting assessment teams to consist of experts in all the disciplines that are relevant to the analysis of social and physical impacts is quite unreasonable, entrusting an interdisciplinary assessment process to individuals who do not know the actual range of disciplinary knowledge is also unreasonable. Teams composed of individuals who are not knowledgeable of social and behavioral disciplines typically fail to recognize the need for collecting and evaluating relevant social data and information.

Although no discipline can be identified as being more important to a comprehensive impact assessment than any other, project managers are well advised to insure a balance of social science and natural science disciplines among assessment team members, as well as a degree of training and experience in these disciplines that is substantially greater than that afforded by introductory academic courses.

INTEGRATIVE TASKS

Integrative tasks of the assessment process include the evaluation of the significance of impacts (Chapter 3) and the identification and evaluation of appropriate mitigation measures (Chapter 19). In general, social impacts that affect relatively few people, that have little effect on public health or economic processes, that arouse little political or other interest or concern, and that do not directly contradict legislated goals and objectives are considered insignificant. This approach to the assessment of social impacts is certainly eminently practical, but it may also be more insensitive to project-mediated impacts on people than it is to impacts on the public exercise of political and legislative power. Environmental impact assessment was not designed as one more tool for implementing majority over minority interests. This process is a tool for making better-informed decisions than would otherwise be made. Thus, any assessment approach that merely perpetuates historical patterns of decision making is, at best, irrelevant and, at worst, contrary to the objectives of environmental impact assessment.

In the evaluation of the significance of social impacts, and prior to the application of the specific criteria suggested in Chapter 3, the assessment team should consider the following factors that might influence the real significance of social impacts:

- the capacity of local individuals and groups to avoid projected impacts that they consider detrimental or otherwise contrary to their interests and desires,
- the long-term commitment to a social and physical environment that future generations have no voice in determining and that will be difficult (if not impossible) to alter should they choose to do so,
- the opportunity of people to use their physical and social environment to obtain their own goals and objectives, as opposed to the enforced forbearance of an environment that physically and socially denies alternative goals and objectives, and
- the sense of justice with respect to the social distribution of project-mediated benefits and risks.

Doubtless many assessment teams will perceive such considerations as falling outside the scope of environmental impact assessment. These concerns are more relevant to social engineering than to the project engineering for which the evaluators have assessment responsibility. Such perceptions signify a basic misunderstanding of the assessment process, which is, after all, predicated on the following precepts:

- As a process of discovery, no one component of the physical or social environment is to be given precedence over any other; the objective is to identify the totality of impacts and their interactions.
- Impact assessment is as concerned with long-term environmental costs and benefits as it is with short-term costs and benefits; this is a necessary consequence of the precept of each generation being considered the trustee of the environment for future generations.
- Jurisdictional limits of agencies providing or acting on proposals for project development do not define the limits of the assessment effort.
- However else the assessment process may be used, it is at least a device for expanding the public's power to influence environmental decision making; it is not intended as a device for limiting or constraining that power.
- No one technical, scientific, or other measure of significance for project-mediated impacts must take priority over any other measure; project development must serve a plurality of environmental interests.
- Even the most sophisticated, balanced, and sensitive assessment of project-mediated impacts is a waste of time if the public does not understand the risks and the benefits that might be realized by project development.

ADDITIONAL CONSIDERATIONS

In addition to managerial aspects of the assessment process, several other factors play key roles in insuring the adequacy of any assessment of social impacts.

Communication

Among all the managerial, technical, and scientific skills that are pertinent to the assessment process, none is more important than communication. Because of too frequent failures in communication, the assessment process often proves inadequate with respect to the identification and evaluation of social impacts.

To communicate is, in the view of many, simply to give information.

Of course, this description of communication is by no means sufficient. Communication is an act that requires, first and foremost, a subjective connection between people, a mutually appreciative relationship that allows the **sharing of meanings** as well as the giving and taking of information. Whether accomplished by oral, written, or visual means, this sharing of meanings makes communication highly personal, even when done quite publicly in the midst of a crowd or over remote separations of distance and time.

The decibel level to be experienced by a homeowner in the near vicinity of a proposed new highway can be calculated. The level of certain air contaminants at various distances downwind from a proposed new power plant, as well as the probability of pedestrian deaths to be expected as a result of projected increases in vehicular traffic associated with a proposed new shopping mall, can be estimated. Having calculated the decibels, estimated the concentrations, and projected accidental mortality, the assessment team can apply certain standards and statistics to determine whether those decibels, concentrations, or mortalities do or do not fall within the appropriate limits. However, what do those same decibels, concentrations, and mortalities mean to the real people who will actually have to live with the risks such numbers imply?

Whatever the meaning the assessment team might ascribe to an impact, other meanings might be ascribed by an informed public that might better reflect the actual social significance of a proposed project. Of course, communication with the public is required for the assessment team to learn what these other meanings might be.

Failures in communication with the public not only deprive the assessment team of knowledge that might be important in evaluating potential impacts, but also of information that is critical for identifying potential impacts in the first place. After all, the local public best knows the local conditions that directly and indirectly influence their behavior, beliefs, values, and concerns; these very conditions might conceivably be affected by project development and thereby result in subsequent social impacts.

Finally, the assessment team must realize that **the assessment process itself is a source of potential impacts on the social environment,** especially with respect to the possible ramifications of poor communication between the assessment team and the public. Certainly, if the assessment process ignores local individuals and groups (neither elicits their views and concerns nor acts on them), or treats the public condescendingly, the resultant frustration, anger, and alienation are impacts as real as any others that might be associated with project development. Like any impact, these impacts may lead to others with possible serious consequences to physical and social components and dynamics.

Field Observations

A wide range of field studies may be implemented in the progress of an assessment. With few exceptions, field studies focus on specific components and dynamics of the physical environment. In many instances, these studies involve significant expenditures of money and person-hours. Field studies devoted to social components and dynamics are seldom considered, with the exception of those that pertain to archaeological or historical sites or to aboriginal peoples. Certainly, the probability of finding assessment personnel out in the general project area examining wildlife habitat and studying their diel, nocturnal, and diurnal migrations is infinitely higher than that of finding them studying human habitats and human behavioral patterns.

The historical proclivity of the assessment community to recruit preferentially the wildlife biologist over the urban anthropologist, the hydrologist over the sociologist, the aquatic biologist over the social psychologist, the soils scientist over the political scientist, and every other specialist trained in natural science disciplines over every other trained in social science disciplines has resulted not only in a dearth of assessment field methodologies and techniques that may inform an assessment team of social impacts, but also in the firm (notwithstanding mistaken) belief that any such field effort is actually inappropriate. The natural science bias that is evident in so many assessment teams means social components and dynamics are often described in terms that would seem to be based on technical disciplinary considerations but actually reflect only a superficial, often misleading, lay understanding.

For example, an assessment team dealing with a neighborhood in the vicinity of a proposed project will most often refer to that neighborhood as a *community* and use such phrases as *community cohesion* to denote an obviously desirable condition or attribute, which then can be used for assessment purposes, as in the identification of a **project-mediated increase or decrease in community cohesion.**

This approach is very logical, but is very likely a totally irrelevant exercise. The fact is that a neighborhood of even densely packed domiciles may not be usefully represented as a community in the usual social science sense, which gives priority to self-sufficiency. Even in the sense of a community as an assemblage of persons and families who achieve and maintain a shared identity through their daily face-to-face interactions, the usual neighborhood in a contemporary pluralistic society probably does not qualify as a community but as a collage of disparate (and often conflicting) interests that simply share a geographic locality. Under such circumstances, what "community cohesiveness" could possibly mean, except as a nostalgic referent to a socially comforting notion, is difficult to imagine.

The primary means of avoiding the use of such concepts is to use field observations as the basis of describing actual social components and dynamics. If the objective is to describe a neighborhood in terms of the actual components and dynamics that give it identity, the field scientist must go out into the neighborhood and study it. After all, this approach is used when a lake, forest, wetland, or other physical resource is the object of interest: use observations of the actual resource to determine the appropriateness of analytical and integrative concepts.

For example, in a particular neighborhood, do people actually spend a significant amount of time interacting with one another on a face-to-face basis? Do particular loci of congregation exist? Do people visit each other in their yards? On porches? On the sidewalk in front of their homes? How are daytime and nighttime patterns of behavior similar and different? What is the rate of resident turnover? Do specific examples of how neighbors cooperate with one another on joint projects related to the neighborhood exist, such as block parties, planting common gardens, and clean-up projects? Where do small children play? Where do older youth congregate, and for what purpose? What do residents say about their neighborhood—the things that please them and the things they don't like? What do local politicians say about the nature of the neighborhood? Local clergy? The police? These are only a few of the questions that people purporting to have a basic understanding of a neighborhood as a social entity should be able to answer. Obtaining the answers is by no means a major effort. In fact, the effort required is most likely substantially less than the effort required to determine the trophic state of a stratified lake, the habitat value of shrub-forest land to local and regional bird species, or the productivity of an estuary—determinations typically made in the progress of assessing impacts on the physical environment. However, such an approach does require the assessment team to be willing to expend an effort and to be cognizant of the importance of an informed overview of the real social components and dynamics that might be affected by project development.

Social Relativism

In the analysis of potential impacts on a detritus-based food web that supports a particular fisheries, for example, that the analyst may be a vegetarian is certainly quite irrelevent. Personal bias is well recognized within the scientific community as a feature that must be avoided and, in physical and social science research, is rigorously controlled. However, in the process of impact assessment, which is definitely not a research effort, controlling personal bias with respect to the assessment of social impacts is often difficult, largely because the assessment is not conducted by research-oriented social scientists, but by persons who have little professional training and experience in the social sciences. Another possibly important reason is that public involvement in the assessment process often results in emotional (if not heated) exchanges between members of the public and the assessment team, a situation that, in the pressurized context of limited time and budget that generally characterizes the assessment process, understandably presents real difficulties for maintaining a composed bias-free attitude with respect to the public. Still another factor is intrinsic to any social analysis: we are all products of our own socialization and necessarily tend to view the rest of the social world in terms of the particular values and beliefs that punctuate our own experience.

The term *social* (or *cultural*) *relativity* is used to denote the need for the social scientist to desist from using his or her own socially derived values and beliefs in the analysis of cultures, societies, and any other social entities that incorporate different values and beliefs. Although a social scientist is not expected to refrain from having or expressing *personal* preferences, in the effort to understand the dynamics of a social group—a group whose behavior might be found personally abhorrent—the social scientist must view those dynamics in terms of the values and beliefs of the members of that group and not in terms of his or her values and beliefs. Additionally, in modern pluralistic societies, the geographic density of different and even conflicting values and beliefs, as well as the behavioral patterns that correlate with those values and beliefs, is so high that any one subgroup or local social entity is unlikely to represent even the semblance of a true microcosm of the total society.

The practical consequences of ignoring social relativism in the assessment of social impacts cannot be overemphasized. For example, an assessment team may consider that direct impacts that result in an increased accessibility of people to people (e.g., removal of intervening forest or other natural obstacles, development of a shopping or recreational area, installation of walkways) are beneficial social impacts. After all, does not accessibility promote the sense of community? In the real world, however, the consequences of removing barriers between people can also result in a dramatic increase in physical conflict. The real issue is whether or not these people want better access to those people and *vice versa*, not what assessment team members, on the basis of their own beliefs and values, think would be nice!

Does the assessment team think that a project-mediated increase in low-skill, low-wage jobs is an economic benefit to a particular locality? Do the people in that locality consider it, instead, an insult? Does the assessment team think that the development of a hazardous waste treatment plant is a long-term benefit regarding the economic stability of a region? Do abutting neighbors see it, instead, as a health threat to their children that is imposed on them by an economically and politically powerful majority that is callously indifferent to the well-being of a powerless minority?

The principle of social relativism requires the assessment team not to accept the values and beliefs of diverse peoples but to recognize them, and not necessarily to sympathize with but certainly to empathize with the differences among people. How else can we know the real social impacts of project development?

PART IV

ISSUES OF
SPECIAL CONCERN

CHAPTER 17

CHEMICAL HAZARD AND RISK ASSESSMENT

In recent years, the rapidly growing concern about chemicals, human health, and environmental quality has resulted in the incorporation of chemical hazard and risk assessment into the environmental impact assessment process.

As part of the more general impact assessment process, chemical hazard and risk assessment is predicated by three basic objectives.

1. **Minimize the potential for locating projects on properties that are contaminated by hazardous chemicals.** The acquisition of contaminated property is likely to result in exorbitant costs for the clean-up of the property as well as for any environmental damage or public health incidents that might be associated with that contamination. Such costs may be the legal responsibility of the current property owner at the time of discovery, whether or not the contamination occurred with the knowledge of, or could in any way be attributed to, that owner. Potential liabilities regarding contaminated property may conceivably extend beyond ownership into leasehold and, under joint-use agreements, may include joint-and-several liability.

Given contemporary economic realities of contaminated property, no responsible agency or private developer can reasonably afford to secure access to real property without having clear evidence that the property does not pose any chemical or other threat to the environment or human health.

2. **Minimize the potential for project-related releases of hazardous chemicals to the environment.** Project-related releases of hazardous chemicals may occur throughtout the various phases of project development, including property management phases prior to construction (e.g., demolition of existing structures that may contain hazardous chemicals), construction (e.g., on-site storage and preparation of materials and supplies, place-

215

ment of contaminated fill), and operation and maintenance phases (e.g., runoff from operational site, application of pesticides).

3. **Minimize the potential for human exposure to hazardous chemicals during all phases of project development.** As with environmental releases, human exposure is of concern in all phases of project development. Exposure must be considered with respect to (a) the public, (b) personnel (including contractors and assessment personnel) involved in the design and construction of the proposed project, and (c) workers (including contractors) employed at the completed facility.

KEY CONCEPTS

Hazard

The hazard of a chemical refers to **potential harm or injury** that may be associated with that chemical. The potential is intrinsic to the chemical and cannot be altered. Any chemical may be associated with several or more hazards.

Of the more than 15 million chemicals known, including naturally occurring and human-made chemicals, approximately 60,000 are in daily commercial use in technologically developed countries. Each of these chemicals may be said to be hazardous because, depending on the degree of exposure, each may result in harm or injury. Certainly some chemicals are more hazardous than others; the more hazardous are most often listed by regulatory agencies as requiring special attention regarding their handling, shipping, and storage.

Chemical hazards may be described as *physical* or *health hazards*. *Hazard class* indicates the specific type of physical or health hazard and is used to categorize a chemical with respect to that hazard. Physical hazards are caused by changes that occur within the chemical and/or its reactants. Appropriate hazard classes include:

1. **Asphyxiant:** A gas or material having vapors that can displace air and cause suffocation
2. **Combustible:** A material that will burn if ignited at a temperature above 100°F and below 200°F
3. **Explosive:** A material that abruptly releases pressure, gas, and heat when subjected to sudden shock, pressure, or high temperature
4. **Pyrophoric:** Any solid or liquid that will spontaneously ignite in air at temperature of 130°F or lower
5. **Organic peroxide:** Any chemical containing carbon and oxygen that is explosively unstable

6. **Oxidizer:** Any chemical that promotes the burning of other substances
7. **Water reactive:** A material that reacts with water to release a flammable or toxic gas
8. **Unstable/reactive:** Any chemical that will spontaneously form different kinds of chemicals that may be flammable or explosive

Health hazards occur when a chemical reacts with living tissue. Appropriate hazards classes include:

1. **Carcinogen:** Any chemical that causes cancer
2. **Corrosive:** Any chemical that burns living tissue on contact
3. **Irritant:** Any noncorrosive chemical that causes temporary itching, soreness, or inflammation of exposed skin, eyes, or mucous membranes
4. **Mutagen:** A chemical that causes changes in the genetic content of a cell
5. **Poison:** A chemical that, in very small amounts (e.g., several drops), can cause death
6. **Toxic:** A chemical that causes damage to tissues or organs and that, at high doses (e.g., several teaspoons or more), can cause death
7. **Sensitizer:** A chemical that causes allergic reactions
8. **Teratogen:** A chemical that causes defects in the developing fetus

Risk

Risk refers not to the intrinsic potential for harm or injury, but to **the probability that an organism will experience that harm or injury.** Generally, risk increases with exposure to the chemical: the greater the exposure, the greater the risk. However, **distinguishing between toxic risks and the risks of other types of harm or injury is important.** Essentially a statistical concept, the risk of experiencing toxicity reflects the fact that **individuals within a biological population demonstrate a range of tolerance with respect to a toxic chemical.** Some individuals are very sensitive to a particular concentration, whereas others may only react to a much higher concentration.

Figure 17.1 depicts an example of a *dose–effect* (or *dose–response*) *relationship* between the dose of a toxic chemical (expressed as weight of the chemical per unit of body weight, e.g., mg/kg) and the probability of a particular effect (e.g., lethality). Dose–effect relationships may be established on the basis of *laboratory experiments* with animals and may also be inferred on the basis of epidemiological studies of humans. In laboratory studies,

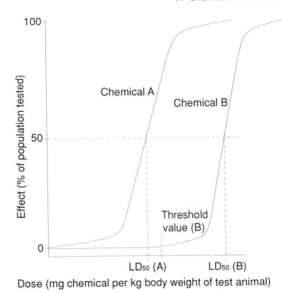

FIGURE 17.1 Dose–effect (response) relationships for two chemicals. Note that some effect (i.e., lethality) is seen for every incremental increase in dose of chemical A, whereas increases in dose of chemical B result only after some threshold value. The percentage of population showing lethal effects may also be read as the probability of death for one organism (e.g., 50% of the population is equivalent to a 0.5 probability for any individual organism).

the experimenter controls the exposure of the test organisms to the toxic chemical. In epidemiological studies, exposures are inferred from information about workplace and other exposures of humans to the chemical of interest, that is, exposures are not controlled by the experimenter but by the life experience of the persons included in an epidemiological study.

As shown in Figure 17.1, the LD_{50} represents the dose at which 50% of the test organisms (of the same species) that are exposed to that dose are expected to die. This statistic essentially states that any one organism exposed to that dose has a 50/50 chance (or 0.5 probability) of dying. If some toxic effect of exposure other than lethality is of interest (e.g., rate of reproduction, loss of hair), an ED_{50} (effective dose) can be determined, in reference to the effect of interest.

LD_{50} and ED_{50} data are based on experiments or epidemiological studies in which the chemical is actually introduced into the organism (e.g., through inhalation, injection); the dose therefore defines that introduced amount. In many instances, the concentration of the toxic chemical in the atmosphere or water in which the test organism lives (i.e., *ambient concentration*) is known, but the amount actually taken into the organism is unknown. In such cases, LC_{50} and EC_{50} are used to denote the lethal concentration for a 0.5 probability of lethality and the effective concentration for a 0.5 probability of some other effect, respectively.

Values of LD_{50} and LC_{50} are very useful for defining relative toxicities. For example, Table 17.1 includes LD_{50} values and commonly used categories of relative toxicity. Although these terms are in general use, LD_{50} values do have important limitations with respect to comparing the toxicity of two or more chemicals. Figure 17.2 depicts the straight line portions of the dose–effect curves for two different chemicals. Note that, although both chemicals have equal LD_{50}s, increasing the dose of one chemical results in a smaller incremental increase in risk than does increasing the dose of the other.

The dose of a chemical received as a result of exposure is of paramount importance with respect to the toxic hazard of a chemical. However, it can be irrelevant with respect to other types of hazard. For example, once allergic to a particular chemical, a person can experience a life-threatening episode on even the most minuscule exposure to that chemical (i.e., allergen). Note that no well-defined relationship exists between the dose of carcinogens, mutagens, and teratogens and the risks of experiencing their respective hazards.

Acute and Chronic Effects

The various effects of chemical exposure may be described as *acute* or *chronic* effects. Acute effects are those that occur very quickly (e.g., minutes, hours, days) after exposure to the causative chemical agent. Asphyxiants, explosives, pyrophorics, organic peroxides, water reactive and unstable/reactive chemicals, corrosives, and poisons typically produce acute effects. Chronic effects are those that occur only long periods of time after

TABLE 17.1 Commonly Used Terms Denoting Relative Toxicity

Relative toxicity	LD_{50} (mg/kg)[a]	Lethal amount[b]	Examples of chemicals[c]
Extremely toxic (poison)	<1	7 drops	Dioxin, botulinum toxin, tetrodotoxin
Highly toxic (poison)	1–50	7 drops–1tsp	Hydrogen cyanide, nickel oxide
Very toxic	50–500	1tsp–1oz	Methylene chloride, phenol
Moderately toxic	500–5000	1oz–1pt	Benzene, chromium chloride
Slightly toxic	>5000	>1pt	Acetone, ethyl alcohol

[a] As tested by the oral route in rats.
[b] Lethal amount for average adult human, based on liquid with density of water.
[c] As tested by various routes in several animal species.

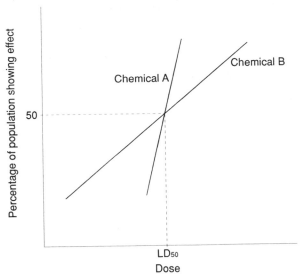

FIGURE 17.2   Dose–effect curves for two chemicals with the same LD_{50}. Note that incremental increases in dose result in greater increases in the percentage of population affected for chemical A than for chemical B.

exposure (e.g., years, decades), including effects of sensitizers and carcinogens. Many lung cancers related to exposure to asbestos and cigarette smoke, for example, develop only after as long as 40 years.

Because chronic effects become apparent only over extended periods of time, they are particularly difficult to relate to a specific exposure to a particular chemical. As a result, our current state of knowledge regarding the potential effects of the more than 60,000 chemicals in daily commerce tends to be much more extensive with respect to their acute effects than their chronic effects. This state of affairs, however, reflects only the methodological difficulties involved in the scientific investigation of chronic effects, and does not imply that chronic effects should be of less concern. In fact, the relative absence of scientific data regarding the chronic effects of exposure to hazardous chemicals (especially at low doses) is sufficient reason to be particularly circumspect regarding the potential for chronic effects.

The phrase "target organ effects" is often used today to specify the particular organs, tissues, cells, and physiologically important systems that are particularly affected by a specific hazardous chemical, regardless of whether the effects are acute or chronic. Target organs typically affected by a wide range of common commercial chemicals include the skin, eye, mucous membranes, respiratory tract, lungs, liver, kidney, reproductive system, and central nervous system.

Dimensions of Exposure

As used earlier, the term "exposure" denotes some measure of the amount of chemical to which an organism is subjected. This measure, for example, "dose" or "concentration," is particularly useful in gauging toxic risks. However, the risk (or probability) of experiencing a toxic effect is not simply a function of dose or concentration. Other factors also influence risk, including age, sex, general state of health, life-style, and any medications (or drugs) a person may use. In some instances, one or more of these factors may dramatically increase the risk of exposure to a hazardous chemical. For example, smoking cigarettes imparts a certain risk to a person of developing lung cancer; so does exposure to asbestos. However, the combination of these two factors results in a substantially greater risk than is imparted by either factor alone.

Another important factor that can directly influence risk is the particular means by which a chemical comes in contact with an organism. Most often referred to as *routes of entry,* these means include (1) inhalation, (2) ingestion, (3) simple contact with skin or eye, (4) absorption of the chemical through the intact skin, eye, or mucous membranes, and (5) puncture (or injection). Not all chemicals will enter the body of an organism through all the possible routes of entry. Most chemicals do, however, enter the body through two or more routes.

The toxicity of many chemicals is greater or lesser (i.e., in terms of the dose required to produce certain effects) depending on the specific chemical and the specific route of entry. For example, a toxic chemical that is ingested will often have a higher LD_{50} (i.e., is less toxic on a per dose basis) than the same chemical injected directly into the bloodstream. Knowing the various routes of entry for different hazardous chemicals is important because, by blocking those routes of entry through the appropriate use of personal protective clothing and equipment (impervious gloves, respirators, etc.), we can effectively prevent exposure and thereby minimize risk.

Environmental Transport and Transformation

Once a chemical enters the environment it is subject to a variety of mechanisms (Chapter 2) that transport it from place to place and from one environmental medium to another (e.g., soil to air, air to water). During transport, a chemical may also undergo transformation because of dynamic physical, chemical, and biological processes (e.g., oxidation, hydrolysis, biotransformation). Some environmentally mediated transformations of chemicals (e.g., biotransformation of certain pesticides) can result in the production of a chemical that is more toxic than the original chemical.

The environmental transport and transformation of a chemical is often referred to as the *environmental fate* of that chemical. Computerized *multimedia environmental models* describe the environmental fate of chemicals and are increasingly available in the scientific literature, governmental publications, and commercial software. These models are important for calculating the probable concentrations of different chemical species in different environmental media, as may be expected as a result of chemical release to the environment. This consideration is necessary in the assessment of the potential exposure of humans and other species to potentially toxic or otherwise hazardous chemicals.

OVERVIEW OF HAZARD AND RISK ASSESSMENT

As shown in Figure 17.3, hazard and risk assessment consists of (1) compiling specific data and information regarding the types of chemical hazard that might be associated with project activity and the various factors influencing the exposure of humans and other species and (2) evaluating these data and information for the likely risks. The types of hazards to be considered depend, of course, on knowledge of the likely chemicals that may be associated with the proposed project site as well as with the various activities associated with all phases of project development.

Chemicals on site prior to the implementation of the project include those chemicals derived from prior use of the site (e.g., pesticides previously deposited to soils, solvents stored in abandoned buildings, metals and organic chemicals absorbed into buried pipes), as well as chemicals transferred to the site from off-site locations (e.g., groundwater contamination from industrial development upstream from the site). Chemicals may also be on site because of accidental spills or illegal dumping of hazardous wastes.

Chemicals associated with project development include chemicals brought onto site, stored and used during construction (e.g., fuels, construction materials and supplies), and likely to be stored, used, or produced during operational and maintenance phases (stock chemicals for industrial production, highway emissions, fertilizers, etc.).

Each chemical should be evaluated with respect to all possible physical and health hazards and potential target organ effects. The exposure assessment must include quantitative data regarding the amounts, transport and transformation of chemicals likely to be associated with the project area prior to, during, and after project development. Determinations should also be made of human and other populations that might come into contact with these chemicals as well as of the specific routes of entry associated with these chemicals. Information regarding potentially exposed human populations should include basic demographic data. The identification of

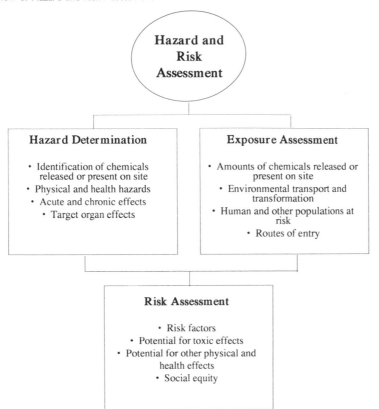

FIGURE 17.3   Key elements of hazard and risk assessment process.

human (and other) populations likely to be exposed should be based on (1) the geographical extent of environmental transport mechanisms (e.g., wind, water) that can transport chemical contaminants to the general public, (2) personnel involved in project development (e.g., construction crews), and (3) personnel employed by the completed facility.

By integrating hazard and exposure information, the risk assessment process determines the potential for acute and chronic toxicity and other health and physical effects on human and other populations. With respect to human populations, health factors such as general health and life-style, which can influence the effects of exposure, must be carefully considered. Such an evaluation requires access to (1) basic demographic data on possibly exposed persons and (2) a review of the technical literature that contains information on factors that are known to be important with respect to specific types of exposure. The distribution of likely health and safety effects within the human population also must be evaluated with respect to the issue of social equity.

Likely effects on nonhuman populations, including floral and faunal populations, should be evaluated for potential subsequent indirect effects on human health (e.g., biomagnification in human food webs), as well as for other effects on the physical and social components and dynamics of the total human environment.

In determining potential risk, the assessment team must work with the fact that, although at least 60,000 chemicals are in daily commerce, health standards are available for only several hundred. Moreover, standards that are available may be strongly biased by current knowledge of the workplace exposure of populations of humans (e.g., relatively healthy, young, caucasian males) who are not necessarily representative of the general human population in terms of their physiological response to chemicals. The assessment team must, therefore, adopt a policy of acting to **minimize exposure** to potential health hazards. The alternative is simply to ignore the overwhelming majority of potential health hazards.

APPLICATIONS TO LOCATION AND PROPERTY ACQUISITION PHASES

Although typically a significant delay occurs between identifying potential alternative locations for project development and the actual acquisition of property, considering these two phases jointly is useful because both involve vital decisions that can greatly affect potential liabilities regarding chemically contaminated real property.

Various techniques are available for identifying the possible contaminated property. However, depending on the size of the property, some of these techniques may not be appropriate. Moreover, prior to the actual legal acquisition of property, very real constraints exist on what can actually be accomplished in terms of site evaluation and analytical testing.

Site Inspection

A preliminary site inspection is basically a visual examination of the property to identify any possible structures, activities, features, or conditions that might indicate historical or potential release of chemicals. The inspection should be carried out by persons who are familiar with the local area in terms of typical types of land use, types of industry, and agriculture. At least one member of the inspection team should be a chemical health and safety officer specifically trained in the recognition and avoidance of chemical hazards.

During the inspection, particular attention should be given to the following:

- present and historical usage of site
- structures, including purpose and type of construction, with special regard for construction materials that present special hazards (e.g., asbestos, lead paint)
- above- and below-ground storage areas possibly used for chemical storage, including tanks (and associated piping), drums, sheds, and stockpiles
- cesspools and sanitary tanks that might contain hazardous chemicals in addition to sewage
- ditches, storm drains, catch basins, and floor drains
- paved surfaces (bitumen and concrete) that may later have to be sounded to insure that they are not hiding pits, tanks, and pipes that might contain hazardous chemicals
- surface water supplies, including open water and wetland areas that might serve as receiving systems for land runoff carrying hazardous chemicals.
- conditions of abutting properties, including structures and activities that could result in the release of chemicals.
- hydrologic gradients in the general area, and land use patterns of up- and down-gradient properties.
- prevailing wind patterns in the general area, and land use patterns of up- and downwind properties
- visual or other evidence (e.g., smells) of previous releases of chemicals, including ground stains, obvious water pollution (e.g., oil scums), stressed vegetation, and trash piles containing pails or other possible chemical containers or contaminated materials including old pipes, construction debris, and tires
- records associated with on-site operation of current facilities, including permits, waste disposal records, chemical inventories, and Material Safety Data Sheets (MSDSs)

In pursuit of this information, beginning a site inspection only after a careful review of aerial photographs and topographic and geological maps is good practice. Conversing with local residents who are likely to have a much better appreciation of local problems that might be related to the release of hazardous chemicals into the environment is also good practice.

Background Records Search

Having conducted a preliminary site inspection, the assessment team should undertake a thorough search of records that might indicate past releases of chemicals that could have contaminated the property. The sources of records of interest will vary, of course, with the particular legal jurisdiction. Examples of the types of information and potential sources that are generally useful follow:

- underground storage tank registration (e.g., local fire department, town or city authorities)
- permits for sewage discharge (e.g., sewer authority, department of public works)
- historical ownership of property (local assessor's office, town clerk)
- water table and artesian wells in general vicinity (local water department or company, department of public works)
- wetland and other natural resources that might have been affected by earlier releases (e.g., local conservation commissions, department of environmental protection, police and fire departments)
- reported releases (police and fire departments, environmental authorities, local emergency response companies and agencies)
- water quality and other reports indicating ambient quality of resources (departments of environmental protection, water resources and wildlife regulatory authorities)

Subsurface Investigation

Conducting a subsurface investigation of property prior to purchase is typically difficult if not actually impossible. However, the importance of a subsurface investigation cannot be overemphasized and every effort should be made to implement one at least as a condition of purchase.

No subsurface investigation should be undertaken until the chemical health and safety officer has devised a comprehensive site safety plan and all field personnel have been trained in its proper implementation, the proper use of field monitors (oxygen meter, combustion meter, organic vapor detector, etc.) and personal protective clothing and equipment. Particularly hazardous conditions noted in the field—such as drums of unknown chemicals, flammable storage sheds, and smells of organic vapors—should be reported immediately to local safety and health officials. All field work should be suspended until appropriate remedial action has been completed.

Techniques typically employed during a subsurface investigation include:

1. **Manual digging/drilling:** Use of shovel and/or auger to determine depth of soil staining; investigate suspicious soils; take limited-depth samples for chemical analysis

2. **Backhoe trenching:** To expose soil profiles to a depth of 10–12 feet; collect soil samples; identify refuse and other burials
3. **Drilling:** Techniques that do not use drilling fluids are preferred; obtain deep soil samples for subsequent laboratory analysis; establish subsurface stratigraphy; establish wells that can be used to monitor up- and down-gradient water quality for documenting on-site water-soluble soil contaminants
4. **Geophysical investigatory techniques:** For locating buried tanks and pipes
5. **Electronic detection of volatile gases:** Used in combination with manual digging, backhoe trenching, and/or drilling

Of course, some of these techniques have practical limitations. For example, manual digging, backhoe trenching, and drilling cannot be applied to an entire property. These methods should be used only when some reason exists to suspect that a particular location might be contaminated. Drilling is most often employed to determine basic groundwater flow patterns and to establish a relatively small number of water quality monitoring wells.

All samples collected for subsequent laboratory analysis should be collected, preserved, and stored according to specific written directions supplied by the laboratory. Parameters to be tested in soil and water samples should minimally include those designated by legal authority as directly relevant to the determination of hazardous wastes and regulated priority pollutants. These substances typically include volatile organics (including halogenated solvents), base/neutral/acid extractables (such as polynuclear aromatic hydrocarbons, phthalate esters, semi-volatile aromatic compounds, and phenolic compounds), pesticides and polychlorinated biphenyls, and inorganic compounds (such as heavy metals and cyanides).

A particular problem associated with the later stages of the property acquisition phase of project development is the demolition of structures on acquired property prior to initiating project-related construction. Ordinarily, the removal of structures and hauling away of wastes is left, all other things being equal, to the lowest bidder. However, the assessment team should evaluate the demolition process as possibly resulting in the generation of a regulated hazardous waste, a situation that would require specifically trained and licensed personnel to remove and properly dispose of demolition wastes.

After all, finding hazardous chemicals stored behind bricked-up walls and under concrete floors of abandoned industrial buildings is not a particularly uncommon event. Asbestos may be included in various components of the structure or its appurtenances. Lead paint may cover wooden surfaces. Creosote may saturate posts and beams. Oils may clog cellar soils. Heavy metals and a wide variety of organic solvents may remain in cesspools and sanitary drains or be absorbed into metal and concrete pipes.

APPLICATIONS TO CONSTRUCTION PHASE

Various events may occur in the progress of construction that raise serious public health concerns regarding the exposure of the public and the construction crews to potentially hazardous chemicals:

- accidental excavation of buried hazardous chemicals, including chemical sludges, covered pits and trenches, pipes, tanks, boxes, bottles, and drums;
- sudden venting of gases through soil disturbed by excavation, blasting, and/or the operation of heavy vehicles;
- release of fugitive (and possibly contaminated) dusts, air emissions (e.g., diesel fumes), and runoff from materials storage and handling areas into the general work area and its surrounding environment;
- construction-mediated changes in hydrological flow that result in contamination by previously unavailable pathways (e.g., downstream underground flow of leachate from a landfill being diverted into areas served by wells);
- fracturing of bedrock or clay separations of water table and artesian aquifers, with possible cross-contamination;
- placement of possibly chemically contaminated borrow in project area; and
- application of paints and preservations to constructed elements, with possible release of fugitive aerosols and liquids into surrounding environment.

Some of these events are certainly accidental and cannot be predicted (e.g., discovery of buried chemicals); some are highly predictable (e.g., fugitive emissions from equipment). To insure the appropriate control of any health hazards, a *health and safety management plan* must be implemented to run concurrently with the construction phase of project development.

The objective of such a plan is to marshal whatever personnel, equipment, and monetary resources may be required to control any health or safety threat to the environment, the public, and the construction crew. This plan should specifically provide for:

- emergency communications with local and regional emergency response agencies and organizations, including fire and police departments, local hospitals, and contractors specializing in the handling and disposal of hazardous chemicals;
- proper use and maintenance of personal protective clothing and equipment;
- emergency evacuation of project-site and surrounding area;
- emergency supplies and materials to be immediately available on site;

- potential signs and symptoms of exposure to hazardous chemicals and appropriate first aid procedures;
- decontamination procedures;
- required health and safety procedures to be implemented during construction-related activities;
- personnel training requirements regarding health and safety procedures; and
- ambient monitoring requirements.

APPLICATIONS TO OPERATIONAL AND MAINTENANCE PHASE

By the time a project has been constructed and enters its operational and maintenance phase, the environmental assessment process has long since been completed. Unlike the early location, acquisition, and construction phases, the assessment team or some other environmental team that might be empowered by project authority has little or no direct control over events that transpire during operational and maintenance phases, except when project operations remain under the control of an operational governmental agency such as a highway or public works department.

Sometimes legally binding agreements are made between private owners of the completed project and local or higher level governmental agencies for the purpose of securing environmental goals and objectives. Such agreements may require specific future actions regarding the influence of operational and maintenance activities on potential chemical exposures. However, in most instances, the operational project simply becomes subject to the normal range of regulatory and other legal constraints at local and national levels, including the requirements specified in ongoing permits and licenses and the common standards of civil and criminal law. Once a project (especially a privately owned facility) becomes operational, the assessment process with respect to the chemical exposure of persons and the environment may appear not only to be complete, but actually irrelevant with respect to day-to-day operations. This is not necessarily the case.

Properly conducted, the assessment process clearly and comprehensively documents the potential operational and maintenance effects of the proposed project regarding chemical exposures. Supposedly, such evaluations are important inputs into the decision-making process of project development. However, the fact that the project becomes operational does not mean that those evaluations suddenly and immediately become moot: they may, in fact, be used in future court proceedings to document that known potential dangers were ignored with serious detriment to environmental quality as well as public health. This crucial aspect of the environmental impact assessment process is often overlooked. **Assessment findings, whatever their disposition in the context of the decision-making process of project development, may become centrally relevant to the decision-making process of subsequent legal proceedings.**

Clearly the operational and maintenance phase of a completed facility may result in environmental and human exposures to the chemicals used, stored, and produced within the property lines of that facility. These exposures may involve persons located within and outside of those property lines. Although assessment team members or the developers of a project may consider themselves to lack any jurisdiction regarding the activities associated with a completed project, they certainly do have responsibility under the rule of *full disclosure*, which is generally applicable to the environmental impact assessment process, to pursue and consider all project-related consequences of chemical exposure.

The assessment team may not be able to control the chemical exposure of workers employed at a completed facility, but such exposures do result from project development and are appropriately identified and evaluated during the impact assessment process.

The assessment team may not be able to alter how those exposures might be distributed inequitably among people on the basis of age, sex, income, ethnicity, or race, but such inequitable distribution may result from project development and are also appropriately identified and evaluated during the impact assessment process.

CHAPTER 18

CUMULATIVE IMPACTS

Many impacts, on a project-by-project basis, prove to be of such small magnitude that they are evaluated to be insignificant. For example, local project-mediated changes in diverse ecological parameters (such as the density of biological populations and the ambient concentrations of chemicals) that are within 10% of ambient values are extremely difficult if not impossible to distinguish from normal fluctuations. Project-mediated impacts of such an order of magnitude are, therefore, generally ignored.

However, evidently low-magnitude effects of individual projects can become highly significant when they are added together over all the projects undertaken in a region over a period of time. For example, various projects, each requiring a small amount of landclearing (e.g., $<<10\%$ of a local forest) may be implemented in a particular area over a period of 20 years. The amount of forest lost to each project may, in effect, be insignificant but the total acreage lost to all project development over that period of time may result in the loss of the entire forest.

Impacts that are essentially insignificant in terms of the individual project that generates them, but may accumulate and become additive in a region over a period time, are referred to as *cumulative impacts*. These effects are of concern because they can easily lead to a "piecemeal degradation" or loss of key environmental components and attributes that, unaddressed, can lead to highly significant long-term (if not irreversible) changes in the total human environment.

Whatever legislative or executive mandate underlies any nationally or internationally defined process of environmental impact assessment, the concern about and the need to address cumulative impacts are at least implicit in the rationale of the assessment process as a tool for making decisions that affect the long-term stability or quality of the environment. This objective is clearly reflected in the well-established precept of each generation acting as the trustee of the environment for succeeding generations. Any assessment process that ignores cumulative impacts specifically negates the value of that assessment process as a primary means of fulfilling this central responsibility.

TYPOLOGY OF CUMULATIVE IMPACTS

A cumulative impact is *simply additive* in the sense that the impact is basically an arithmetic summation of the same type of incremental impact, as when the total forest acreage lost becomes the sum of the incremental losses associated with each successive developmental project in an area. Simply additive cumulative impacts include those that result from gradual increases in the ambient concentrations of potentially toxic or injurious chemicals in air, water, and soil and finally result in statistically significant environmental and public health risks, such as those associated with the phenomena of acid rain (which can result in important reductions in forest and aquatic productivity) and the contamination of drinking water supplies with heavy metals and other hazardous chemicals. These impacts also include those that result from gradual decreases in resources, including wildlife resources (such as habitats and species), human recreational and educational resources, and scenic, historical, and archaeological resources.

In other instances, a cumulative impact may be greater than the simple arithmetic summation of the incremental contributions made by subsequent projects. For example, a population of large mammals (e.g., deer) may be successively reduced in proportion to the amount of critical habitat removed by project development. However, the total population may be eradicated long before the entire critical habitat is removed, because population density is not simply a function of habitat availability, but also depends on other factors including a minimal density required for successful reproduction. Similarly, the total forest acreage lost to successive project developments is often greater than the sum of all incremental losses in forested acres associated with successive clear-cutting, because the number of trees actually felled is a function not only of purposeful clear-cutting but also of windthrow, a phenomenon associated with the abrupt exposure of previously protected, shallow-rooted trees to wind. An impact that is greater than the arithmetic total of the incremental contributions made by successive projects may be described as *synergistic*.

Simply additive and synergistic cumulative impacts may mediate their own consequences. For example, successive development of various kinds may result in the incremental reduction of overstory in a region. At some point, the vegetative denudation results in catastrophic wind erosion, with the result that regional nutrient-rich topsoils are swept away. The remaining soil stratum becomes unfit for agriculture and locally dependent human populations suffer starvation. This phenomenon is most commonly referred to as *desertification*. In this particular scenario of desertification, the cumulative impact is the regionally significant disappearance of overstory. Its direct consequence is wind erosion; the consequence of wind erosion is the failure of agriculture; the consequence of the failure of agriculture is starvation.

Certainly one cannot attribute the starvation of total human populations to the loss of overstory that might be associated with any specific local project or activity. Instead, the cumulative influence of a range of activities (and other environmental factors) over space and time triggers a causal chain that can lead to starvation. Perhaps this consideration is the most useful to keep in mind regarding cumulative impacts: not only are they gradual accumulations of known environmental degradations and insults, but they also are potential triggers to unforeseen environmental consequences. Such global consequences cannot be readily determined if we focus only on the seemingly innocuous individual degradations and insults from which they emerge.

As a potential trigger to unforeseen and possibly calamitous consequences, note that a cumulative impact may have not only physical but also social dimensions. The significance of the gradual accumulation of contributory impacts is not necessarily measured in terms of critical concentrations, critical densities, or critical masses of physical, chemical, and biological attributes, but in terms of critical limits on social and psychological thresholds and phenomena as well.

In addition to desertification, global warming and the loss of the globally protecting ozone layer are examples of potential cumulative impacts of human activity that are of increasing concern. In addition, many other examples of cumulative impacts, although not deserving of global notoriety, are potentially significant to local and regional human populations.

Example 18.1   Over a period of 20 years, a suburban residential community of about 6 square miles becomes hemmed in by two parallel highways as well as several secondary roads that interconnect the highways. Subdivision development associated with the highways does not directly intrude into the community, but does include two large malls and the usual fast-food and gas station development in close proximity to the connectors and highways.

The noise associated with vehicular traffic not only increases with respect to the highways, but also changes character as a greater proportion of the mid- to late-evening highway traffic becomes heavy-duty commercial trucks that service two major metropolitan areas east and west of the residential community, as well as the local malls. The noise, in terms of decibels and pitch, as well as the increase in local traffic along community roads, is a significant nuisance to community residents who increasingly perceive local traffic as an escalating safety hazard, the persistent noise as a physical intrusion into their former peaceful privacy, and the increase in the nonresidential population as a threat to their seclusion and their security.

These common sentiments directly influence increasing numbers of families to sell their homes. However, the features that cause these families to move out of the community are the same features that discourage families from buying into the community, with the result that residences gradually give way to small retail and professional service businesses. Within another 10 years, the residential community is effectively displaced; even a local school building is converted into a privately owned commercial industry.

Note that, in this example, various simply additive cumulative impacts are caused by successive project developments, including impacts related to noise levels, traffic congestion, and the density of nonresident populations. At some point, all these impacts synergize in terms of human psychology, with consequent secondary impacts on the demography and economics of the affected community, none of which is likely to be triggered by any individual development implemented over the 20-year period.

Note also that, in this example, broad social trends that are not directly related to a specific project at the time of its inception (e.g., the increasing density of heavy duty trucks on highways) are suggested nonetheless to influence the eventual manifestation of a project-related impact significantly (i.e., highway noise). This situation underscores the need for an assessment team to understand that a project-related impact should not be considered inherent in the project itself but an ultimate derivation of the project in context. **Changing attributes and conditions of the physical and the social environment must be evaluated for their influence on the manifestation of the consequences of individual projects and their cumulative impacts.**

Example 18.2   A state water resource authority conducts a boat-rental operation in one of its controlled watersheds. The purpose of this operation is to provide (for a fee) small motorized boats to fishermen who are restricted to one warm water impoundment that is not part of the regulated water supply. The generated money is used for a wide range of maintenance activities within the watershed. No money is used for restocking the impoundment; the objective is to utilize a self-sustaining warm water fishery.

Over a period of several decades, the boat service becomes increasingly popular, showing a substantial increase in boat rentals every several years up to the physical capacity of the agency's ability to store and maintain boats. However, a severe regional drought occurs at the beginning of the fourth decade of operations. Within a few seasons, the water level of the impoundment falls sufficiently low that nuisance vegetation is able to establish itself throughout the impoundment. Because the vegetation entangles lines and lures, fishing steadily decreases and finally ceases for the duration of a remaining 5-year drought.

On resumption of the normal hydrological regime, fishing is expected to increase to its previous levels. However, even after several seasons, it is evident that it will not because the relative abundance of non-sport fish species is dramatically greater than the abundance of sport fish species. The boat-rental service is eventually discontinued because of lack of interest.

In this example, the implementation of the boat-rental service directly results in an increase in the number of hours of fishing expended in the impoundment. From a limnological perspective, the direct impact of the boat service is the introduction of a significant predator into the aquatic ecosystem who does not prey equally on all warm water species, but selectively harvests specific species. However, in actual practice, pole and line fishing typically results in significant losses of nontarget species that, having been hooked, are thrown back and, having been injured, are at higher risk of disease and natural predation. Thus, for every individual of the target species harvested by the human predator, several (e.g., 2–6) individuals of the nontarget species die. If the nontarget species is competitive with the target species (for food or habitat), human predation will favor proliferation of the target species.

In light of these considerations, the sustained incremental increase in fishing over the observed period of time might reflect the gradual development of a mature positive feedback mechanism by which the activity of fishermen enhances the availability of the very fish species desired by those fishermen. In essence, then, a possible cumulative impact of the boat-rental service might be the establishment of human predation as a dominant factor in the relative distribution of warm water fish species.

The drought is, of course, an independent factor that results in the removal of the long-established dominant human predation and thus allows natural competition among sport and non-sport fish species to reemerge as the key determinant of the relative population densities of the various species. In this sense, the very success of the boat-rental service, as measured by the cumulative emergence of human predation as a prime determinant of the aquatic ecology, results in the utter collapse of the environmental condition (availability of sport fish) the service was designed to exploit. As in Ex. 18.1, this result is abruptly made manifest by a changing environmental condition (i.e., the drought) that is totally independent of the cumulative direct impact of the boat-rental service.

BUREAUCRATIC CONSTRAINTS ON THE ASSESSMENT OF CUMULATIVE IMPACTS

Historically, impact assessment has been conducted in association with the decision-making process of *project development*, the planning and implementation of a specific plan with a relatively narrow-focused objective, as exemplified in such phrases as "highway project," "power plant project," "water diversion project," and "new community development project." In some cases, impact assessment is associated with *prescribed activities,* such as "logging or conversion of forest land to other land use within the catchment area of reservoirs used for municipal water supply," "construction of hospitals with outfall into beachfronts used for recreational purposes," and "construction of ports." Impact assessment has also been associated with *programmatic activities,* which include a wide range of alternative policy approaches to similar objectives, such as "control of nuisance aquatic vegetation" and "management of commercially important shellfish." Whatever the terminology used to describe the bureaucratic basis of impact assessment, **the environmental impact assessment process is (with few exceptions) necessarily constrained by a defined set of actions undertaken within the strictly limited jurisdictional authority of specific governmental agencies having operational (as opposed to environmental policy) responsibility.** Although the history of impact assessment is coincident with the relatively recent development of governmental agencies that have broad environmental responsibility and authority, such agencies are typically not the agencies (or, in some instances, the private developers) who initiate and conduct the assessment process. Instead, they establish guidelines and regulations pertaining to the conduct of that process.

Insofar as impact assessment remains the primary responsibility of operationally focused governmental agencies (transportation, agriculture, energy, flood control, etc.), assuming that the assessment process will give adequate attention to the issue of cumulative impacts—which, after all, are not singularly pertinent to any particular type of project or activity but most often reflect the totality of human effort at local, regional, national, and even global levels—is unrealistic. A broadly empowered agency with specific and general environmental responsibilities that extend well beyond the relatively narrow operational interests of all other governmental agencies would be well positioned to exercise the type of comprehensive authority required to insure the proper assessment of cumulative impacts. However, such a role would quickly establish that agency as an extremely powerful arbiter among politically resourceful contenders—a not quite impossible but surely politically improbable event under current circumstances.

PRACTICAL APPROACHES

Notwithstanding the development of national and international authorities that purposely look beyond individual projects to the cumulative consequences of larger assemblages of human activities, the assessment of potential cumulative impacts is still within the province (limited as it may be) of the typical assessment team that has environmental responsibility regarding individual project development. From a practical perspective, the assessment of cumulative impacts as well as of any impacts rests on an appropriate description of the *current environmental setting,* which is defined as the condition of the physical and social environments prior to the projected implementation of any of the proposed project alternatives. This before-and-after comparison of the environment is the heart of the assessment process. After all, how can one assess environmental "changes" caused by project development unless one first defines the basis for recognizing those changes?

In defining the baseline environmental conditions in comparison with which project-mediated changes can be identified, the assessment team should carefully consider the following points.

1. Baseline environmental conditions are inclusive of physical and social components, attributes, and dynamic processes. They are also inclusive of any factors that may influence or be influenced by specific components, attributes, and dynamic processes.

2. Although describing the baseline environmental conditions of the immediate project area is certainly necessary, understanding that environmental components, attributes, and dynamic process are not constrained by the arbitrary property lines that project engineers draw on maps is of vital importance. The geographic extent beyond project boundaries that must be considered is precisely the extent required to describe fully the potential dynamic interactions among project-related actions and activities and the environment.

3. In addition to insuring that baseline environmental conditions are described in a geographic extent that is sufficiently large to account for all project–environment interactions, the assessment team must also insure that a period of time is considered that is long enough to encompass the multitude of potential project-mediated impacts. At a minimum, this period of time should be at least as long as the projected lifetime of the project.

Although the description of baseline conditions in terms of these requirements is essential for the assessment of any type of impact, the specific relevance of these requirements to the assessment of cumulative impacts should be emphasized.

First, the diminution of a particular type of resource (e.g., high quality water for indigenous fish species) may reasonably be considered not only a factor that may influence the evaluation of a specific impact of a particular project, but also a factor that is itself influenced by that evaluation.

For example, if a proposed project may add contamination to a stream previously contaminated with industrial effluent, the assessment team might consider the project's impact on that stream to be insignificant. The stream is already toxic to fish, and additional contamination will not affect them. Does this judgment not verify and give acceptance to the previous historical trend? If so, what is the likely future of currently contaminated streams, except a continuously cumulative degradation? On the other hand, what if the assessment team uses the historical trend as a justification for evaluating project-related contamination as unacceptable? If this were the decision, the assessment would tend to validate a policy contrary to historical precedent that, in future assessments of future projects, could result in a cumulative improvement in water resources.

This example underscores the importance of the full awareness of the assessment team of executive, legislative, and regulatory goals and objectives that most often go well beyond the purview of an individual project and can be used by the assessment team as the basis for evaluating the significance of cumulative impacts.

Second, extending the geographic extent beyond the limited project area forces the assessment team to consider seemingly remote implications of project development. These considerations not only are necessary to assess the consequences of such specific high-profile impacts as fugitive air emissions and noise, but also foster an assessment of high- and low-profile impacts in terms of a broad geographical context. This perspective is necessary for identifying and evaluating many types of cumulative impacts. Filling in a small wetland area, for example, may appear to be of negligible concern when the general project area is replete with numerous, much more extensive wetlands. Considering the general project area in its broader geographic context, however, the assessment team may realize that even that small wetland is a relatively rare resource and should be given special consideration.

Third, the assessment of the significance of impacts may be greatly influenced by an expanded appreciation of time, which underlies the precept of each generation being the trustee of the environment for future generations.

Of particular importance is the consideration that the assessment of cumulative impacts depends not only on (1) the simple accumulation of incremental impacts over time, but also on (2) changing environmental conditions and factors and (3) changing knowledge and values.

The incremental contributions of contaminants in highway runoff to local watersheds should be evaluated not in terms of the highway's projected 50-year lifetime alone, but also with respect to other changes in the environment expected within that same period, for example, changing traffic volume and nature of cargo, especially with respect to industrial chemicals. Consideration should also be given to current regulatory trends, such as decreasing maximum contaminant levels that might require future reexamination of the potential toxicity of highway runoff.

A FINAL NOTE

No doubt the assessment of cumulative impacts is an intellectually demanding, technically difficult, and bureaucratically distasteful task. Assessment teams are prone to argue that they have neither the authority nor the wherewithal to undertake this task. Even if it were attempted, it would be an utter waste of time because such a task requires extensive predictions of future events that cannot be predicted with any degree of accuracy.

With respect to the question of authority, one may readily reply that the consideration of cumulative impacts, even when not specifically required by legal mandate, is absolutely intrinsic to any assessment process that purports to provide decision makers with information needed for the balanced consideration of environmental alternatives. Although real constraints are imposed on the assessment process regarding the consideration of cumulative impacts, constraints do not obviate the responsibility to do so.

The question of predictability is appropriately addressed by reminding assessment teams that the impact assessment process is, like the decision-making process it serves, a marshalling of the information, insight, and judgments that have been thoughtfully examined. In this regard, inadequate assessments are far more likely to result from the refusal to consider aspects that are not blatantly obvious than from the common inability to foresee the future with comforting precision.

CHAPTER 19

MITIGATION

In the earliest literature relating to impact assessment, mitigation was understood to be inclusive of the following objectives and actions:

- to avoid an adverse impact altogether by not taking a certain action,
- to minimize adverse impacts by limiting the degree or magnitude of the action,
- to rectify an adverse impact by repairing, rehabilitating, or restoring the affected environment,
- to reduce or eliminate an adverse impact over time by appropriate techniques of maintenance or preservation, and
- to compensate for an adverse impact by replacing or providing substitute resources or environments.

Assessors quickly realized that such objectives and actions were too limited because they focused entirely on making adverse impacts less severe. Mitigation, therefore, became inclusive not only of any means for avoiding, reducing, or compensating for adverse impacts but also of any means for promoting or increasing the environmentally beneficial impacts of project development.

General objectives commonly adopted for the minimization of adverse impacts and the maximization of beneficial impacts include the following:

1. **Avoidance:** Avoiding projects or activities that could result in adverse impacts; avoiding certain types of resources or areas considered to be environmentally sensitive

This approach, which is most appropriate to the earliest phases of project planning, is generally considered the most important of mitigation measures. The success of this approach depends on the timely availability of environmental data and information as well as on a consensus regarding significant environmental issues.

  2. **Preservation:** Preventing any future actions that might adversely affect an environmental resource or attribute

This goal is typically achieved by extending legal jurisdiction beyond the immediate needs of project development to selected resources. However, many operational governmental agencies are prohibited from taking land that is not specifically required for project development.

  3. **Minimization:** Limiting the degree, extent, magnitude, or duration of adverse impacts

This approach is probably the most common, and requires careful consideration of a wide range of engineering and project management techniques and methods.

  4. **Rehabilitation:** Rectifying adverse impacts by repairing or enhancing the affected resources

Many ecosystems may be rehabilitated to enhance selected attributes, such as biological productivity and wildlife habitat. This approach is usually appropriate when previous development or contamination has resulted in the significant diminution of the environmental functions and attributes of a particular resource.

  5. **Restoration:** Rectifying adverse impacts by restoring the affected resources to an earlier (and possibly more stable and productive) state

Restoration is, in essence, the extreme of rehabilitation. This method typically requires an extensive and intensive engineering of a selected resource to achieve what might be considered "pristine" conditions.

  6. **Replacement:** Compensating for the loss of an environmental resource at one location with the creation or protection of that same type of resource at another location

Widely practiced, this approach is typically coupled with the objective of preservation and, in such instances, most often involves transference of legal ownership of the replaced resource to some agency or organization for the express purpose of preserving it from any future development.

  7. **Improvement:** Enhancing the capability of an existing resource with respect to its environmental functions

Like minimization, improvement requires consideration of a wide

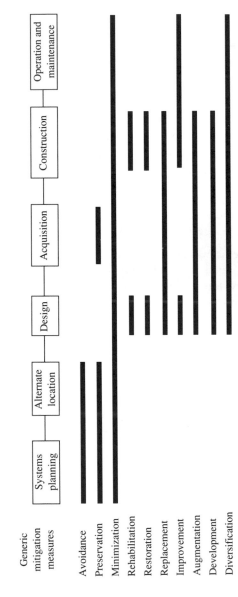

FIGURE 19.1 Relevance of various mitigation measures to different phases of project development.

range of engineering and project management techniques that may be implemented to enhance a particular environmental function or attribute.

8. **Augmentation:** Increasing the area or size of an existing environmental resource

A form of improvement, augmentation narrowly focuses on the geographic (areal) or morphometric (depth, areal configuration) attributes of an aquatic, terrestrial, or wetland resource.

9. **Development:** Creating specific environmental resources in an area in which they are currently absent

Widely exemplified by the development of borrow pits as new wetland resources, this approach has been increasingly applied to terrestrial and aquatic ecosystems. However, the creation of a new resource requires careful consideration of the interactions between the new resource and its environs to insure that the resource will be self-sustaining.

10. **Diversification:** Increasing the mixture or diversity of habitats, species, or other environmental resources in a circumscribed area

Although diversification may include the creation of a new resource, it typically involves the introduction of new habitat opportunities and/or new floral or faunal species. Because of the complexities involved (both scientific and political) in introducing new species, this approach is typically restricted to extensive development projects with appropriately large budgets of time and money that can be devoted to detailed environmental studies.

As shown in Figure 19.1, the various objectives of mitigation are appropriately implemented in different phases of project development. A successful approach to mitigation therefore requires that the impact assessment process, which identifies and evaluates the impacts to be mitigated, be well advanced, even at the earliest systems planning phase. Simply, the longer the assessment process is delayed with respect to initial planning, the more opportunities for mitigation are lost.

DEVELOPING A COMPREHENSIVE MITIGATION PROGRAM

The identification of appropriate mitigation objectives and the implementation of specific measures to achieve those objectives requires a *programmatic approach* that will insure detailed consideration of the following questions:

- Which impacts can and should be mitigated?
- What specific methods are available for achieving the objectives of mitigation?

- What factors can influence the successful implementation of specific methods?

Significance of Impacts

The mere fact that an impact may occur does not mean that the assessment team should immediately consider how to mitigate that impact. After all, the key objective of impact assessment is not simply to identify impacts but to evaluate those impacts, to determine which shall be considered significant and which shall be considered insignificant.

No doubt many mitigation efforts have been implemented in instances in which the targeted impacts were essentially insignificant, sometimes quite consciously to appease political or other interests that might otherwise, without any environmental basis, impede project development. Sometimes such mitigation is done because of a mistaken belief that all impacts are equally important, and that project development can and should result in a "net zero impact" on the environment.

Any act of mitigation is essentially an expenditure of public money to achieve specific environmental objectives. Moreover, a mitigation measure may result in unforeseen environmental consequences. Therefore, the assessment team must insure that **the expenditure of public money to implement mitigation be made only with respect to significant impacts and that such expenditures do not result in consequences that are contrary to environmental objectives.**

Cause–Effect Relationships

Any impact may have several contributing causes. For example, the loss of fisheries from a stream may result from (1) abrasive suspended particles that are released during land clearing operations and abrade sensitive gills, (2) sedimentation of particles in project area runoff onto developing eggs, (3) development of oxygen deficiency in the stream because of project-mediated loading of the stream with organic materials, (4) deposition of clay-sized nonabrasive particles on gills, (5) increase in temperature of water because of removal of overhanging vegetation, (6) removal of detrital food supply as a result of dredging, and (7) toxic effects of project-related runoff, discharge, or spillage of chemicals.

Assuming that the loss of the fisheries is considered significant, and that this impact must be mitigated, the assessment of alternative mitigation measures must be conducted with precise consideration of the specific causal pathways leading to the impact. For example, should the mitigation focus on control of suspended particles released during project construction, or should it concentrate instead on the removal of food supply, the reduction in dissolved oxygen, the introduction of toxic chemicals, or other potential causes?

Of course, before addressing these questions, the assessment team will have to come to grips with other key questions.

- What is the relative contribution of each possible cause to the overall impact?
- Must mitigation be implemented to prevent the impact totally, or might reducing the magnitude of the impact be sufficient?
- Is it reasonable to allow the impact to occur and, after the completion of the project, restock the stream with the same or other comparable species?

In some cases, an impact attributed to project development may also be attributable to other environmental factors that, even in the absence of the project, would exert their effect. For example, long-term subdivision development in a local area may be responsible for the gradual increase in eutrophication (Chapter 7) of a surface water supply because of nutrient releases from septic systems as well as from the actions of swimmers (urination directly into the water) and boaters (stirring up of bottom muds). Assessors may determine that a proposed project will also contribute to the eutrophication through the release of nutrient-laden soils during land clearing or through rapid fertilization techniques applied during the landscaping of the project.

In such a case, the impact of project development is not the eutrophication of the surface supply, which is already underway because of other non-project-related factors. The real impact of the proposed project is to shorten the period in which a specific degree of eutrophication will be achieved. Does a 1-year, 5-year, or even 10-year delay in the inevitable decline in water quality, bought at the price of mitigating project-related contributions to an ongoing process, actually warrant the cost of that mitigation?

Structural and Nonstructural Methods

Structural methods of mitigation are those that require major investments of money to design, construct, and maintain structural features of the proposed project. Such features may typically be composed of steel or concrete and may involve significant amounts of earthwork. They include such features as underpasses and overpasses designed for the migration of animals under and over highways, fish ladders and elevators, and specially designed impoundments and wetlands. Structural methods also include fencing designed to control wildlife access to rights-of-way, as well as extensive terrestrial, aquatic, and wetland habitats specifically designed, constructed, and landscaped to provide food supply, shelter, and isolation.

Nonstructural methods of mitigation include managerial protocols, such as the staging of project-related activities, for example, dredging to avoid environmentally significant events, like the migration of anadromous fish, and the storage of excavated soils so they can be replaced without mixing soil horizons. Nonstructural methods might also include the use of relatively low cost materials that are essential to mitigating certain types of impacts, such as hay bales or silt screens to control the particulate loading of streams, and riprap to maintain slope stability.

Once the assessment team has determined significant impacts, the team is best advised to consider the appropriateness of nonstructural means of mitigation first, for the following reasons:

1. Nonstructural means do not require major expenditures of money. Therefore, if assessors subsequently determine that the selected means does not, in fact, result in desired objectives, sufficient budget will remain to attempt other approaches.

2. Nonstructural means typically do not require extensive maintenance; thus, their effectiveness does not depend on long-term maintenance budgets, which may or may not be feasible.

3. Nonstructural means do not require the placement of significant structures in the environment that, regardless of their intended effect, may have unforeseen consequences that cannot be easily rectified without the removal of those structures.

Structural means of mitigation should only be considered once assessors determine that the intended mitigation cannot be achieved using nonstructural means.

EVALUATION OF STRUCTURAL AND NONSTRUCTURAL ALTERNATIVES

In evaluating the various measures that might appropriately mitigate a significant impact, assessors must consider a wide range of site- and project-specific factors (Table 19.1). The following examples are illustrative of the practical relevance of some of these factors.

Example 19.1 A highway is constructed with a relatively wide median strip. Members of the design team are interested in designing the median strip to meet a variety of objectives, including aesthetic values,

TABLE 19.1 Examples of Criteria for Evaluating the Appropriateness of Proposed Mitigation Measures

Regional factors acting against success of the proposed measure
Compatibility with environmental values and objectives
Level or measure of success of proposed measure
Secondary impacts of proposed mitigation measure
Longevity of effects of proposed measure
Potential for interference with operation of proposed or other projects
Potential consequences if proposed mitigation measure fails
Cost (including implementation and maintenance)
Maintenance requirements of proposed measure
Jurisdictional authority to implement proposed measure
Previous experience with proposed measure
Long-term monitoring effort required
Potential for future alterations to improve success of proposed measure
Potential for vandalism

vehicular safety, and, if possible, wildlife values. The decision is made to plant the median strip with a long-flowering shrub, which will meet aesthetic objectives as well as safety objectives because of the impact-absorbent capacity of woody shrubs. The team further proposes to select a long-flowering woody shrub that produces berries, which may have food value for local bird populations.

The project is completed and, over a period of several years, receives much public praise because of the beauty of the flowers and an excellent record of vehicular safety. The public and the wildlife biologists are also pleased that the shrub's berries do attract local bird populations, which adds to the aesthetic value of the median strip. Overlooked, however, is the fact that, in periods of drought, the berries of the selected shrub species ferment. This oversight becomes clearly manifest during an actual drought as hordes of intoxicated birds weave unpredictably among the traffic lanes of the highway. They cause so many vehicular accidents that the median strip must be uprooted entirely.

Example 19.2 A small military-related facility is constructed in a desert, in an area known to be used by the desert tortoise for its various migrations. Fencing is used in an effort to guide migrating tortoises safely through the facility. However, the selected mesh size proves to be too large; many of the tortoises are found to have entangled their heads and necks in the mesh, with consequent suffocation and fatal cuts.

Example 19.3 In an effort to improve the safety of a secondary road, a state department of public works realigns the road to remove a particularly dangerous curve. After construction of the new alignment, the curve, no longer needed, is abandoned in its paved condition. In an effort to reduce costs, no attempt is made to excavate it and return it to a forested area. Access to it is blocked at either end with large boulders.

Within 2 years, local authorities suddenly discover that the abandoned roadway curve has become an ideal drinking hideaway and "lovers' lane" for local teenagers after four 16-year-olds are killed in a late night highway accident after attending an extended beer party held there. Several pregnancies among teenagers can also be attributed to the easily accessible and strongly sheltered privacy of the abandoned road.

Example 19.4   An extensive tree-planting project is undertaken as part of an urban beautification effort in a temperate region. Tens of thousands of dollars are spent in the purchase and planting of a variety of trees specifically selected for the beauty of their foliage and their high resistance to various types of urban air contaminants. Overlooked, however, is the fact that the street lighting in the areas of these plantings use mercury-vapor lamps. These lamps have the particular effect of prolonging the growing season of the selected tree species. During the first hard freeze of autumn, all these hardy, temperate trees die because their physiology is still behaving as in summer.

Example 19.5   In an effort to mitigate the roadkill of deer in a remote area, an expensive underpass is constructed to serve as a means for deer to migrate safely from one side of a highway to the other. Fencing on either side is designed to funnel the deer—first into the underpass and then away from the underpass into the shrub- and forest-land on the other side of the highway. The underpass is outfitted with remotely controlled cameras to document its use as a safe passage.

Within several seasons, local people determine that although roadkill is significantly reduced by the safe passage provided by the underpass, so is the density of the local deer populations. Local hunters, who are certainly no less clever than wildlife biologists, consider the underpass an excellent device for funneling the deer into their gunsights.

Example 19.6   A local developer of a golf course agrees with the demand of local authorities to modify an existing 5-acre wetland to make it more suitable for wood-duck and to deed the modified wetland to the state. The modified wetland, which will abut the golf course, is intended as environmental compensation for a 1-acre wetland that must be filled to accommodate the golf course.

Modification of the 5-acre wetland requires a minor diversion of stream flow and the construction of several earthen berms to retain water at the desired depth which, in many places, is up to 3 feet deeper than the original wetland. During the first summer after construction of the project, two small children wander from their nearby yard into the modified swamp and drown.

As evidenced by some of the preceding examples, the failures of some mitigation measures are clearly attributable to illegal or at least inappropriate human behavior. In such instances, the argument may be made that, after all, the abandonment of a piece of roadway in the woods (Ex. 19.3) does not cause teenage drinking or teenage pregnancy, and the design of a deer underpass (Ex. 19.5) does not cause the illicit use of that structure as a shooting gallery. However, illegal or at least inappropriate human behavior is, in fact, real data about a real condition in the human environment. We must consider those data in the appropriation of the public's money and resources for the achievement of mitigation. Ultimately, for whatever reasons, a mitigative measure either achieves its objective or does not.

Most of the preceding examples emphasize that the implementation of a mitigative measure can often result in unforeseen consequences. Sometimes, the reason these impacts are unforeseen is our lack of precise environmental knowledge and understanding. However, sometimes (Ex. 19.2, 19.6) the reason is simply our failure to conduct an adequate assessment of facts and conditions that are easily discernible, if they are considered.

The failure to consider adequately those facts and conditions of the real environment as they pertain to the objectives of mitigation is most often a direct consequence of the following situations.

1. In many jurisdictions, the legal requirement for conducting impact assessment pertains only to governmental authorities, whether at national, federal, state, provincial, regional, or local levels. It does not pertain to the private sector. In such instances, governmental authorities must work within strictly defined jurisdictional limits.

A highway department, for example, typically has authority with respect to activities and structures within a precisely delineated right-of-way. What happens outside that right-of-way is not within the scope of the highway department. In accord with its environmental responsibility, the department may implement a well-conceived measure of environmental mitigation within its right-of-way, for example, a highway underpass to promote the safe migration of animals. However, that department has no jurisdictional authority with respect to privately owned lands on either side, the actual use of which by private landowners may undo the department's mitigation efforts.

In this context, governmental agencies disregard what they cannot control. However, to insure the cost-effectiveness of mitigation, the evaluation of any proposed mitigation measure must include careful consideration of how the limits of jurisdictional authority might influence the efficacy of the measure.

2. Limited jurisdictional authority, coupled with geographic isolation, often means that agency personnel responsible for mitigation operate with little knowledge or understanding of the mitigation efforts undertaken

by their colleagues in other geographical areas or by their counterparts in different agencies. Thus, in one jurisdiction, personnel in one agency may undertake the design of extensive structures to accomplish a mitigation that someone else has achieved by a less extensive or even more effective means. Generally speaking, case studies of actual mitigation measures, including design features, as well as an analysis of key factors that influence the success or failure of the measure, are relatively difficult to come by.

3. The vast majority of any mitigative effort is typically undertaken by individuals who are trained and educated in engineering and natural science disciplines. The design and implementation of any structural or even nonstructural mitigation measures absolutely requires specific knowledge and understanding of the "mechanics" by which a project is designed, constructed, and operated.

Such a mechanical preoccupation with mitigation does lessen the probability that human behavior and psychology will be adequately addressed with respect to their potentially significant contribution to the failure or the success of any planned mitigation or to the development of unforeseen consequences.

GUIDELINES

1. Each proposed mitigation measure should be considered an activity like any other undertaken in the course of project development. In other words, a proposed mitigation measure should be assessed for all its potential impacts on the social and physical components and dynamics of the environment, just as every other project-related activity (e.g., land clearing, excavation) should be assessed. Mitigation should not be assessed only with respect to its intended objective. This approach will help insure that unforeseen secondary impacts of proposed mitigation measures are duly identified and evaluated for their significance with respect to the total human environment.

2. Alternative mitigation measures should be considered as soon as potentially significant impacts are identified so appropriate mitigation may be implemented, if possible, during early locational and design phases of project development. The more quickly mitigation measures are identified in project development, the more likely nonstructural approaches will prove to be appropriate and the more likely sufficient time and money will be available to conduct a thorough evaluation of structural mitigation measures.

3. In evaluating the desirability and probable efficacy of any proposed mitigation measure, the interdisciplinary team should evaluate all physical and social factors (e.g., Table 19.1) that may influence its successful implementation. Special attention should be given to reviewing any docu-

mentation of the measure's successes or failures in previous mitigation efforts. Preference should be given to those measures that have been previously implemented. Proposed new measures should only be considered when mitigation objectives cannot be obtained by the use of previously tested measures.

In projects requiring the review by a variety of governmental and other agencies and organizations, a proposed mitigation measure should be implemented only after a consensus has been achieved among these agencies and organizations on the overall appropriateness of the measure, its specific causal relevance to the impact to which it is addressed, and the level or degree of mitigation to be achieved by the measure.

Any cost–benefit analysis conducted in the progress of impact assessment should include consideration of the distributional cost (Chapter 14) of each mitigation measure, the distributional benefits to be realized by each measure, and the length of time over which costs and benefits will be realized.

4. Once implemented, the mitigation measure should be monitored for its success with respect to the mitigation objective and also for any unforeseen consequences. This documentation is vital to insuring the economic long-term achievement of environmental objectives in subsequent developmental projects by providing the assessment community with tested and proven methods.

SUGGESTED READINGS

CHAPTER 1

Gray, Andrew. 1991. *Between the Spice of Life and the Melting Pot: Biodiversity, Conservation and Its Impact on Indigenous Peoples*. Copenhagen: International Work Group for Indigenous Affairs

Hindmarsh, R. A., (eds.). 1988. *Papers on Assessing the Social Impacts of Development*. Brisbane, Australia: Griffith University Press

Rossini, F. A., and A. L. Porter. 1983. *Integrated Impact Assessment*. Boulder, Colorado: Westview Press

Vennum, Thomas, Jr. 1988. *Wild Rice and the Ojibway People*. St. Paul, Minnesota: Minnesota Historical Society Press

Westman, Walter E. 1985. *Ecology, Impact Assessment, and Environmental Planning*. New York: John Wiley and Sons

The World Bank. 1982. *The Environment, Public Health, and Human Ecology: Considerations for Economic Development*. Washington D.C.: The World Bank

CHAPTER 2

Howard, Philip H., *et al.* 1991. *Handbook of Environmental Degradation Rates*. Chelsea, Michigan: Lewis Publishers

Knox, Robert C. 1992. *Subsurface Transport and Fate Processes*. Chelsea, Michigan: Lewis Publishers

Larson, Richard A. 1991. *Reaction Mechanisms in Environmental Organic Chemistry*. Chelsea, Michigan: Lewis Publishers

Mackay, Donald. 1991. *Multimedia Environmental Models: The Fugacity Approach*. Chelsea, Michigan: Lewis Publishers

Manahan, Stanley E. 1991. *Environmental Chemistry*. Chelsea, Michigan: Lewis Publishers

Meerschaert, Mark. 1993. *Mathematical Modeling*. Orlando, Florida: Academic Press

Moriarty, F. 1988. *Ecotoxicology: The Study of Pollutants in Ecosystems*. New York: Academic Press

CHAPTER 3

Bartell, S. M., R. H. Gardner, and R. V. O'Neill. *Ecological Risk Estimation*. Chelsea, Michigan: Lewis Publishers

Beanlands, G. E., and P. N. Duinker. 1983. *An Ecological Framework for Environmental Impact Assessment in Canada.* Ottawa, Canada: Federal Environmental Assessment Review Office

Knowles, Porter, C. 1992. *Fundamentals of Environmental Science and Technology.* Rockville, Maryland: Government Institutes

Michael, Gene Y. 1991. *Environmental Data Bases: Design, Implementation, and Maintenance.* Chelsea, Michigan: Lewis Publishers

World Health Organization and Centre for Environmental Management and Planning. 1992. *Environmental and Health Impact Assessment of Development Projects: A Handbook for Practitioners.* New York: Elsevier Applied Science

CHAPTER 4

American Public Health Association. 1985 et seq. *Standard Methods for the Examination of Water and Waste Water.* Washington, D.C.: American Public Health Association

American Society for Testing and Materials. 1984. *Standard Practice for Evaluating Environmental Fate Models of Chemicals.* Philadelphia, Pennsylvania: American Society for Testing and Materials

Barnthouse, L. W., and G. W. Suter II (eds.). 1986. *User's Manual for Ecological Risk Assessment,* ORNL-6251. Oak Ridge, Tennessee: Oak Ridge National Laboratory

Hunsacker, C., and D. E. Carpenter (eds.). 1990. *Ecological Indicators for the Environmental Monitoring and Assessment Program.* Washington, D.C.: U.S. Environmental Protection Agency, Office of Research and Development

International Agency for Research on Cancer. 1986. *Evaluation of Methods for Assessing Human Health Hazards from Drinking Water,* IARC Internal Technical Report No. 86/001. Lyon, France: International Agency for Research on Cancer

Knapp, C. M., D. R. Marmorek, J. P. Baker, K. W. Thomton, J. M. Klopatek, and D. F. Charles. 1990. *The Indicator Development Strategy for the Environmental Monitoring Assessment Program.* Washington, D.C.: U.S. Environmental Protection Agency, Office of Research and Development

National Research Council. 1989. *Biological Markers of Air-Pollution Stress and Damage in Forests.* Washington, D.C.: National Academy Press

CHAPTER 5

Connell, D. W. 1989. *Bioaccumulation of Xenobiotic Compounds.* Boca Raton, Florida: CRC Press

Craig, P. J. 1980. *The Natural Environment and the Biogeochemical Cycles.* New York: Springer-Verlag

Heling, E. 1990. *Sediments and Environmental Geochemistry: Selected Aspects and Case Histories.* New York: Springer-Verlag

Jorgensen, S. E., and I. Johnsen. 1989. *Principles of Environmental Science and Technology.* Amsterdam: Elsevier Science Publishers

McIntosh, R. P. 1985. *The Background of Ecology: Concept and Theory.* Cambridge: Cambridge University Press

Paul, E. A., and F. E. Clark. 1988. *Soil Microbiology and Biochemistry.* San Diego, California: Academic Press

van Leeuwen, Merman P., and Jacques Buffle. 1993. *Environmental Particles* (2 vols.). Chelsea, Michigan: Lewis Publishers

World Health Organization. 1987. *Health and Safety Component of Environmental Impact Assessment*, Environmental Health Series No. 15. Copenhagen: World Health Organization

CHAPTER 6

Advanced Studies in Science, Technology and Public Policy. 1982. *A Study of Ways to Improve the Scientific Content and Methodology of Environmental Impact Analysis. Final Report to the National Science Foundation.* Grant PRA-79-10014. Bloomington, Indiana: School of Public and Environmental Affairs, Indiana University
Bartlett, Robert (ed.). 1989. *Policy through Impact Assessment: Institutionalized Analysis as a Policy Strategy.* New York: Greenwood Press
Bergman, H. L., R. A. Kimerle, and A. W. Maki. (eds.). 1986. *Environmental Hazard Assessment of Effluents.* New York: Pergamon Press
Hart, Stuart L., Gordon A. Enk, and William F. Hornick. (eds.). 1984. *Improving Impact Assessment: Increasing the Relevance and Utilization of Technical and Scientific Information.* Boulder, Colorado: Westview Press
U.S. Environmental Protection Agency. 1992. *Environmental Equity: Reducing Risks for All Communities,* EPA-230-DR-92-002. Washington, D.C.: U.S. Environmental Protection Agency
The World Bank. 1984. *Environmental Guidelines.* Washington, D.C.: The World Bank

CHAPTER 7

Adams, S. M. (ed.). 1990. *Biological Indicators of Stress in Fish.* Bethesda, Maryland: American Fisheries Society
Baker, J. P., D. P. Bernard, S. W. Christensen, M. J. Sale, J. Freda, K. Heltcher, P. Scanlon, P. Stokes, G. Suter, and W. Warren-Hicks. 1990. *Biological Effects of Changes in Surface Water Acid-Base Chemistry,* State-of-Science/Technology Report 13. Washington, D.C.: National Acid Precipitation Program
Barret, G. W., and R. Rosenberg (eds.). 1981. *Stress Effects on Natural Ecosystems.* New York: John Wiley and Sons
Clark, K. E., F. A. P. C. Gobas, and D. Mackay. 1990. Model of organic chemical uptake and clearance by fish from food and water. *Environ. Sci. Technol.* 24:1203–1213
Connell, Des W. 1992. *Chemical Dynamics in Freshwater Ecosystems.* Chelsea, Michigan: Lewis Publishers
Hellawell, J. M. 1986. *Biological Indicators of Freshwater Pollution and Environmental Management.* London: Elsevier Applied Science Publishers
Kennedy, V. S. (ed.). 1984. *The Estuary as a Filter.* Orlando, Florida: Academic Press
Oceans and Coastal Areas Programme Activity Center of UNEP. 1990. *An Approach to Environmental Impact Assessment for Projects Affecting the Coastal and Marine Environment.* Nairobi, Kenya: United Nations Environment Programme
Tenore, K. R., and B. C. Coull (eds.). 1980. *Marine Benthic Dynamics.* Columbia: South Carolina: University of South Carolina Press
U.S. Environmental Protection Agency. 1990. *Biological Criteria: National Program Guidance for Surface Waters,* EPA-440/5-90-004. Washington, D.C.: U.S. Environmental Protection Agency, Office of Water

CHAPTER 8

Bailey, R. G. 1975. *Ecoregions of the United States.* Ogden, Utah: U.S. Forest Service

Carter, M. R. (ed.). 1993. *Soil Sampling and Methods of Analysis.* Chelsea, Michigan: Lewis Publishers

Hagen, Hohn (ed.). 1992. *Ecology and Conservation of Neotropical Migrant Land Birds: Proceedings of a Symposium.* Washington, D.C.: Smithsonian Institution

Hof, John. 1992. *Coactive Forest Management.* Orlando, Florida: Academic Press

Hunter, Mark D., *et al.* (eds.). 1992. *Effects of Resource Distribution on Animal–Plant Interactions.* Orlando, Florida: Academic Press

Marini-Bettolo, G. B. 1986. *Chemical Events in the Atmosphere and Their Impact on the Environment.* New York: Elsevier Science Publishing Company

Noble, R. D., J. L. Martin and K. F. Jensen. 1989. *Air Pollution Effects on Vegetation, Including Forest Ecosystems.* Broomall, Pennsylvania: Northeastern Forest Experiment Station

Sigal, L. L., and G. W. Suter II. 1987. Evaluation of methods for determining adverse impacts of air pollution on terrestrial ecosystems. *Environ. Mgmt.* 11:675–694

Smith, W. H. 1990. *Air Pollution and Forests.* New York: Springer-Verlag

Somerville, L., and C. Walker (eds.). 1990. *Pesticide Effects on Terrestrial Wildlife.* London: Taylor and Francis

CHAPTER 9

Adamus, Paul R. 1982 *et seq. A Method for Wetland Functional Assessment* (2 vols.). Washington, D.C.: U.S. Department of Transportation, Federal Highway Administration

Beccasio, Angelo D., *et al.* 1982. *Gulf Coast Ecological Inventory User's Guide and Information Base.* Washington, D.C.: U.S. Fish and Wildlife Service, Biological Services Program

Good, Ralph E., Dennis F. Whigham, and Robert L. Simpson. (eds.). 1978. *Freshwater Wetlands: Ecological Processes and Management Potential.* New York: Academic Press

Greeson, Phillip E., John R. Clark, and Judith E. Clark. 1979. *Wetland Functions and Values: The State of Our Understanding.* Minneapolis, Minnesota: American Resources Association

Lonard, Robert I., E. J. Clairain Jr., R. T. Huffman, J. W. Hardy, L. D. Brown, P. E. Ballard, and J. W. Watts. 1981. *Analysis of Methodologies Used for the Assessment of Wetlands Values.* Washington, D.C.: U.S. Water Resources Council

Marble, Anne D. 1992. *A Guide to Wetland Functional Design.* Chelsea, Michigan: Lewis Publishers

U.S. Department of the Interior, Fish and Wildlife Service. 1982. *Wetland Community Profiles* (6 vols.). Washington, D.C.: U.S. Fish and Wildlife Service, Office of Biological Services

CHAPTER 10

American Society of Testing and Materials. 1990 *et seq.* Standard guide for conducting 10-day static sediment toxicity tests with marine and estuarine amphipods. *Annual Book of ASTM Standards* 11.04:1052-1075

Keith, Lawrence H. (ed.). 1987. *Principles of Environmental Sampling.* Washington, D.C.: American Chemical Society

Mackay, Donald, *et al.* 1993. *Illustrated Handbook of Physical-Chemical Properties and Environmental Fate for Organic Chemicals.* Chelsea, Michigan: Lewis Publishers

Mount, D. I., and L. Anderson-Carnahan. 1988. *Methods for Aquatic Toxicity Identification Evaluations: Phase I; Toxicity Characterization Procedures*, EPA-600/3-88-034. Duluth, Minnesota, U.S. Environmental Protection Agency

Ramamoorthy, S., and E. Baddaloo. 1991. *Evaluation of Environmental Data for Regulator and Impact Assessment*. New York: Elsevier Science Publishing Company

Sandhu, Shahbeg S., *et al.* (eds.). 1990. *In Situ Evaluation of Biological Hazards of Environmental Pollutants*. New York: Plenum

Vouk, V. B., G. C. Butler, A. C. Upton, D. V. Park, and S. C. Asher. 1987. *Methods for Assessing the Effects of Mixtures of Chemicals*. Chichester, England: John Wiley and Sons

CHAPTER 11

Connerton, Paul. 1989. *How Societies Remember*. Cambridge: Cambridge University Press

Odum, H. T. 1971. *Environment, Power & Society*. New York: Wiley-Interscience

Marcus, George E., and Michael M. J. Fischer. 1986. *Anthropology as Cultural Critique: An Experimental Moment in the Human Sciences*. Chicago, Illinois: University of Chicago Press

Ritzer, George (ed.). 1990. *Frontiers of Social Theory: The New Synthesis*. New York: Columbia University Press

Sztompka, Piotr. 1991. *Society in Action: The Theory of Social Becoming*. Chicago, Illinois: University of Chicago Press

Turnbull, Colin. 1983. *The Human Cycle*. New York: Simon & Shuster

CHAPTER 12

Branch, Kristi, Douglas A. Hooper, James Thompson, and James Creighton. 1984. *Guide to Social Assessment: A Framework for Assessing Social Change*. Boulder, Colorado: Westview Press

Breakwell, Glynis M. (ed.). 1992. *Social Psychology of Identity and the Self Concept*. Orlando, Florida: Academic Press

Carley, Michael J., and Eduardo S. Bostelo. 1984. *Social Impact Assessment and Monitoring: A Guide to the Literature*. Boulder, Colorado: Westview Press

Goffman, E. 1959. *The Presentation of Self in Everyday Life*. Garden City, New York: Doubleday

Mackie, Diane M., and David L. Hamilton. 1993. *Affect, Cognition, and Stereotyping*. Orlando, Florida: Academic Press

Schwing, R. C., and W. A. Albers, Jr. (eds.). 1980. *Societal Risk Assessment: How Safe Is Safe Enough?* New York: Plenum Press

CHAPTER 13

Agency for Toxic Substances and Disease Registry. 1988. *The Nature and Extent of Lead Poisoning in Children in the United States: A Report to Congress*. Atlanta, Georgia: Centers for Disease Control

Birley, M. H. 1991. *Forecasting Potential Vector Borne Disease Problems of Irrigation Schemes*. Copenhagen: World Health Organization

Calabrese, Edward J. 1985. *Toxic Susceptibility: Male/Female Differences.* New York: John Wiley and Sons

Calabrese, Edward J. 1992. *Biological Effects of Low Level Exposures to Chemicals and Radiation.* Chelsea, Michigan: Lewis Publishers

Groopman, J. D. and P. Skipper (eds.). 1991. *Molecular Dosimetry and Human Cancer: Analytical, Epidemiological and Social Considerations.* Boca Raton, Florida: CRC Press

Maher, Edward. 1992. *Environmental Health.* Cambridge, Massachusetts: Harvard University Press

Paustenbach, D. J. (ed.). 1990. *The Risk Assessment of Environmental and Human Health Hazards: A Textbook of Case Studies.* New York: John Wiley and Sons

Polednak, Anthony P. 1989. *Racial and Ethnic Differences in Disease.* Oxford: Oxford University Press

World Health Organization. 1987. *Health and Safety Component of Environmental Impact Assessment,* Environmental Health Series No. 15. Copenhagen: World Health Organization

World Health Organization Regional Office for Europe. 1983. *Environmental Health Impact Assessment of Irrigated Agricultural Development Projects: Guidelines and Recommendations.* Copenhagen: World Health Organization

Zakrzewski, Sigmund F. 1991. *Principles of Environmental Toxicology.* Washington, D.C.: American Chemical Society

CHAPTER 14

Canter, Larry W., Samuel F. Atkinson, and F. Larry Leistrite. 1985. *Impact of Growth: A Guide for Socio-economic Impact Assessment and Planning.* Chelsea, Michigan: Lewis Publishers

Costanza, R. 1991. *Ecological Economics: The Science and Management of Sustainability.* New York: Columbia University Press

Folmer H., and E. van Ierland. 1989. *Valuation Methods and Policy Making in Environmental Economics.* New York: Elsevier Science Publishing Company

Galbraith, J. K. 1973. *Economics and the Public Purpose.* Boston, Massachusetts: Houghton Mifflin

Nealey, S. M., and E. B. Liebow (eds.). 1975. *Cost Benefit Analysis and Water Pollution Policy.* Washington, D.C.: The Urban Institute

Swedberg, Richard. 1990. *Economics and Sociology: Redefining Their Boundaries.* Princeton, New Jersey: Princeton University Press

Zukin, Sharon, and Paul DiMaggio (eds.). 1990. *Structures of Capital: The Social Organization of the Economy.* Cambridge: Cambridge University Press

CHAPTER 15

Atkinson, Paul. 1990. *The Ethnographic Imagination: Textual Constructions of Reality.* New York: Routledge

Bodley, John H. (ed.). 1988. *Tribal Peoples and Development Issues: A Global Overview.* Mountain View, California: Mayfield Publishers

Gedicks, Al. 1993. *The New Resource Wars: Native and Environmental Struggles against Multinational Corporations.* Boston, Massachusetts: South End Press

Geertz, C. 1973. *The Interpretation of Cultures: Selected Essays.* New York: Basic Books

Geisler, Charles C., R. Green, D. Usner, and P. West. (eds.). 1982. *Indian SIA: The Social Impact Assessment of Rapid Resource Development on Native Peoples,* Monograph No. 3. Ann Arbor, Michigan: University of Michigan, Natural Resources Sociology Research Laboratory

Horowitz, Donald L. 1985. *Ethnic Groups in Conflict.* Berkeley, California: University of California Press

Polesetsky, Matthew (ed.). 1991. *Global Resources: Opposing Viewpoints.* San Diego, California: Greenhaven Press

Smith, George S. 1991. *Protecting the Past.* Boca Raton, Florida: CRC Press

West, Patrick C., and Steven R. Breching (eds.). 1991. *Resident Peoples and National Parks: Social Dilemmas and Strategies in International Conservation.* Tuscon, Arizona: University of Arizona Press

CHAPTER 16

Covello, Vincent T., *et al.* (eds.). 1989. *Effective Risk Communication: The Role and Responsibility of Government and Nongovernment Organizations.* New York: Plenum Press

Finsterbusch, Kurt, *et al.* (eds.). 1990. *Methods for Social Analysis in Developing Countries.* Boulder, Colorado: Westview Press

Nothstein, G. Z. 1984. *The Environmental Liability Handbook for Property Transfer and Financing.* Chelsea, Michigan: Lewis Publishers

Ryding, Sven-Olof. 1992. *Environmental Management Handbook.* Chelsea, Michigan: Lewis Publishers

The World Bank. 1984. *Occupational Health and Safety Guidelines.* Washington, D.C.: The World Bank

CHAPTER 17

Calabrese, Edward J., and Elaina M. Kenyon. 1990. *Air Toxics and Risk Assessment.* Boca Raton, Florida: CRC Press

Finkel, A. 1990. *Confronting Uncertainty in Risk Management.* Washington, D.C.: Resources for the Future

Goldsmith, John R. 1986. *Environmental Epidemiology: Epidemiological Investigation of Community Environmental Health Problems.* Boca Raton, Florida: CRC Press

Hallenbeck, W. H., and K. M. Cunningham. 1986. *Quantitative Risk Assessment for Environmental and Occupational Health.* Chelsea, Michigan: Lewis Publishers

Hess, Kathleen. 1993. *Environmental Site Assessment, Phase 1: A Basic Guide.* Chelsea, Michigan: Lewis Pulishers

Organization for Economic Cooperation and Development. 1989. *Compendium of Environmental Exposure Assessment Methods for Chemicals,* Environmental Monograph No. 27. Paris: Organization for Economic Cooperation and Development

Rappaport, S. M. and Thomas J. Smith (eds.). 1991. *Exposure Assessment for Epidemiology and Hazard Control.* Chelsea, Michigan: Lewis Publishers

Sara, Martin N. 1993. *Standard Handbook of Site Assessment for Solid and Hazardous Waste Facilities.* Chelsea, Michigan: Lewis Publishers

Seip, Hans M., and Anders B. Heiberg (eds.). 1989. *Risk Management of Chemicals in the Environment.* New York: Plenum Press

Suter, Glenn W., II 1993. *Ecological Risk Assessment.* Chelsea, Michigan: Lewis Publishers
Wilson, Albert R. 1992. *Environmental Risk: Evaluation and Finance in Real Estate.* Chelsea, Michigan: Lewis Publishers

CHAPTER 18

Beanlands, G. E., W. J. Erkmann, G. H. Orians, J. O'Riordan, D. Polcansky, M. H. Sadar, and B. Sadler. 1985. *Cumulative Environmental Effects: A Binational Perspective,* No. EN 106-2/1985. Ottawa, Canada: Minister of Supply and Services
Cada, G. F., and C. T. Hunsaker. 1990. Cumulative Impacts of Hydropower Development: Reaching a Watershed in Impact Assessment. *Environ. Profess.* 12:2–8
Canadian Minister of Supply and Services. 1986. *Proceedings of the Workshop on Cumulative Environmental Effects: Setting the Stage.* Ottawa, Canada: Minister of Supply and Services
Carley, Michael. 1984. *Cumulative Socioeconomic Monitoring: Issues and Indicators for Canada's Beaufort Region.* Ottawa, Canada: Department of Indian Affairs and Northern Development
Gosselink, J. G., L. C. Lee and T. Muir. 1990. *Ecological Processes and Cumulative Impacts Illustrated by Bottomland Hardwood Ecosystems.* Chelsea, Michigan: Lewis Publishers
Hamilton, R. S. (ed.). 1991. *Highway Pollution.* New York: Elsevier Science Publishing Company
Hunsaker, C. T., R. L. Graham, G. W. Suter II, R. V. O'Neill, L. W. Barnthouse, and R. H. Gardner. 1990. Assessing ecological risk on a regional scale. *Environ. Mgmt.* 14:325–332
Kroon, M. (ed.). 1991. *Freight Transport and the Environment.* New York: Elsevier Science Publishing Company
Peterson, E. G., Y.-H. Chan, M. M. Peterson, G. A. Constable, R. B. Caton, C. S. Davis, R. R. Wallace, and G. A. Yarranton. 1987. *Cumulative Effects Assessment in Canada: An Agenda for Action and Research,* Cat. No. EN 106-7/1987E. Ottawa, Canada: Minister of Supply and Services
Suter, G. W. II. 1990. Endpoints for regional ecological risk assessments. *Environ. Mgmt.* 14:9–23

CHAPTER 19

Barcelona, Michael, *et al.* 1990. *Contamination of Ground Water: Prevention, Assessment, Restoration.* Park Ridge, New Jersey: Noyes Data Corporation
Cowherd, C. *et al.* 1990. *Control of Fugitive and Hazardous Dusts.* Park Ridge, New Jersey: Noyes Data Corporation
Marsh, William M. 1991. *Landscape Planning Environmental Applications.* New York: John Wiley and Sons
Nyer, Evan K. 1992. *Practical Techniques for Groundwater and Soil Remediation.* Chelsea, Michigan: Lewis Publishers
Turner, M. G. and R. H. Gardner (eds.). 1991. *Quantitative Methods in Landscape Ecology.* New York: Springer-Verlag

Wang, Charleston C. K. 1987. *How to Manage Workplace Derived Hazards and Avoid Liability*. Park Ridge, New Jersey: Noyes Data Corporation
Zonneveld, I. S., and R. T. T. Forman (eds.). 1990. *Changing Landscapes: An Ecological Perspective*. New York: Springer-Verlag

INDEX